Statistics for Business
Fourth Edition

Statistics for Business

Fourth Edition

DEREK GREGORY

HAROLD WARD

ALAN BRADSHAW

McGRAW-HILL BOOK COMPANY

London · New York · St Louis · San Francisco · Auckland · Bogotá Caracas · Lisbon · Madrid · Mexico · Milan · Montreal · New Delhi Panama · Paris · San Juan · São Paulo · Singapore · Sydney Tokyo · Toronto

Published by McGRAW-HILL Book Company Europe
Shoppenhangers Road, Maidenhead, Berkshire, SL6 2QL, England
Telephone 0628 23432   Fax 0628 770224

British Library Cataloguing in Publication Data
Gregory, Derek
   Statistics for Business. – 4Rev.ed
   I. Title   II. Ward, Harold
   III. Bradshaw, Alan Robert
   519.5
   ISBN 0–07–707610–9

Library of Congress Cataloging-in-Publication Data
Gregory, Derek.
   Statistics for business/Derek Gregory, Harold Ward, Alan
Bradshaw. — 4th ed.
     p.   cm.
   Includes bibliographical references and index.
   ISBN 0–07–707610–9 :
   1. Statistics.   2. Commercial statistics.   I. Ward, Harold,
   II. Bradshaw, Alan,   III. Title.
   QA276.12.G73   1993
   519.5′02465–dc20
                                   92–43484
                                      CIP

1234 CL 9543

Typeset by TecSet Ltd, Wallington, Surrey
Printed and bound in Great Britain by Clays Ltd., St. Ives plc.

To
Jennifer and Joyce

Contents

Preface to the 4th Edition

Since the last edition of this textbook the micro-computing revolution has happened. No statistics text can afford to ignore this and, to reflect current practice in the teaching and use of business statistics, we have added computing sections and chapters where we think they will be most useful to students already familiar with the basic keyboard.

This has been done, we hope, in the 'user friendly' way we have always adopted when writing this book, drawing a broad line between avoiding knitted brows and insulting the intelligence of the reader. The use of seven different packages gives the book versatility and we have been careful to concentrate on popular and well-regarded software. At the same time it is perfectly possible for those who do not use computers to study from the familiar Gregory and Ward updated text which has led so many students through the thicket of formulae and figures to success in the practice and examination of this subject.

We are grateful to DRS Data & Research Services plc for permission to reproduce the EDPAC answer sheet in Chapter 2 and, as in previous prefaces, we repeat our thanks to the many teachers, students, firms and public institutions who have contributed to this text. We hope that students in universities, colleges and schools, those working in distant learning situations, and business-men and women in industry and commerce, will continue to find this book as useful as the many thousands of their predecessors in the UK and overseas.

Our thanks are due to Robert Gregory for his painstaking work in revising the index to this new edition.

Derek Gregory
Harold Ward
Alan Bradshaw

1

Introduction to statistics

Statistics deal with numbers, but not simply with counting. The science of statistics is concerned with the *comparison* of numbers. **Statistics defined**

For example, the statement that exports were a certain figure in 1989 means nothing to us if we have no idea of what the figure was in 1987 or in 1988. But the statement that a particular person is 2 m 10 cm tall may provoke a certain surprise because we can mentally compare this figure with the normal heights that we know.

Figures in statistics are not significant unless we can make comparisons. This means that the figures we compare must:

1. Refer to the same group of things.
2. Be of the same kind.
3. Be in the same units.
4. Be measured to the same degree of accuracy.
   Note. Although accuracy is always desirable, it is often not necessary and, more often, not possible. Here we mean, for example, that the figures we compare should be carried to the same number of decimal places.

An example might be as follows.

'In 1989 total exports from the UK had increased by more than 13 per cent on the 1988 figure, when we exported less than £500 m of electrical machinery to the European Community countries.'

In this statement, the errors which hinder true comparison are:

Error	1988	1989
Group	European Community countries	All countries
Kind	Electrical machinery	All goods
Units	£ million	Percentages
Accuracy	To nearest £100 m	To whole units

The statement, therefore, is nearly meaningless and probably misleading.

Meaning of statistics

Although the word statistics is used loosely to mean collections of figures on different subjects, e.g. Department of Employment statistics, the word also applies to the statistical methods which we use. We can define the two meanings of the word statistics as:

1. A collection of comparable figures in a subject or in a connected group.
2. The methods of treating or processing a series of figures.

This book deals with both meanings; the first meaning in Chapter 2, and the second meaning in nearly all the subsequent chapters.

Growth of statistics

Statistical method was first used in the nineteenth century when the first serious attempt to gather figures referring to the population was made in the first census of 1801. These figures, known as *vital statistics*, refer to births, marriages, and deaths. But official collections of figures, dealing with persons, cattle, wealth, land, etc., go back to the beginnings of recorded history.

Nobody collects figures on a large scale just for the fun of it. Therefore, every attempt—from Roman times, to the Domesday Book of 1086, to the first census of 1801, to the latest government report—has been made with a definite purpose in mind. Early kings and rulers usually wanted to know the military strength which could be raised from the population under their control. But a simpler reason than this was that kings, like everyone else, must be fed and housed. Yet kings give services rather than produce goods and, because their services have no market value, they must tax their subjects to raise the necessary finance to provide for the court and to carry on the business of government. Whom should they tax? Obviously, those able to pay. Kings must, therefore, compile figures of the wealth and incomes of their subjects by some method, bearing in mind that one of the first charges on the tax revenue raised will be the cost of the survey and the cost of collection.

Because early societies were extremely poor by our standards, and because statistics are extremely expensive to collect, only the absolutely essential figures were collected: for example, figures of wealth, incomes, customs, fighting men, and so on, and these figures were usually sketchy and not very accurate.

In the nineteenth century, in the Western world, industry and commerce grew so rapidly and to such proportions that historians gave the name of 'industrial revolution' to the fantastic increase in output and the new methods of production. As countries became richer, so their governments were able to raise larger tax revenues to pay for the heavier jobs they had to do. It became more urgent for the government to discover the extent of the changes in the pattern of wealth which resulted from the industrial development. Because the tremendous revolution in production and trade brought distress and injustice in its wake, there had to be similar revolutions in the sanitary, welfare, educational and medical fields. Doctors are at a great disadvantage if they lack records of the causes of death, and of how outbreaks of disease are distributed throughout the country. Educationists cannot plan future school building without a knowledge of how many children exist in a country and what their ages are—and even what the birth-rate might be expected to be in the near future. Statistics are required as a *basis for action* in all fields wherever information can be measured.

A similar expansion occurred in the business field. Another feature of nineteenth-century growth was an increase in the average size of the business unit in most industries. It is possible to run a one-man business (or a 'small master' business) by keeping the figures in one's head. But a large concern, probably operating on mass-production techniques, will certainly be split into different departments, almost like government departments. Some of these departments might deal with labour relations, sales, market research, quality control, stock control, production, design, superannuation—depending on the size and nature of the firm. Large firms, each with their own individual 'Department of Employment' and 'Treasury' exist on the basis of figures. The wealth of such firms often encourages them to make seemingly lavish expenditures on the kinds of figures they collect, although no firm is large enough to undertake surveys on the same scale as the Government. A firm may launch a statistical enquiry costing thousands of pounds to discover how many housewives still boil their sheets. This is of limited national interest and would never be the subject of a government inquiry, yet it may be absolutely vital information to a firm manufacturing detergents.

Present and future statistics

Statistics cannot run a business or a government. Nor can the study of statistics do more than provide a few suggestions or offer a few pointers as to a firm's, or a government's, future behaviour. Although production may in the future be carried on by an automated process, nobody will ever be able to produce the right *decisions* for five years ahead simply from figures fed into a statistical computing machine.

Statistics can be used as a tool of management to tell managers what has happened in the past and (if the figures are collected quickly enough) what is happening now, thus providing a surer basis for decision. Every day, managers are trying to predict consumer demand and the Government is seeking to understand the changing pattern of our balance of payments by the use of statistics. But the further ahead 'forecasts' are made, on the basis of past and present statistics, the more completely are our forecasts likely to be turned upside-down by the course of events. Basically, people and events are so unpredictable and are subject to so many uncontrollable influences, that forecasting is dangerous unless the period of projection into the future is very short indeed.

Today, the output of statistics is almost an industry in its own right with an output which is an ever-growing, and even menacing, flood. Statistics are hurled at the public by newspapers, periodicals, radio, films, television, handouts, posters on hoardings and through the post. The public are increasingly asked to complete official returns and questionnaires, and to give opinions in the street. Form-filling has become a national pastime, but although we may voluntarily labour for hours on a football coupon, or willingly complete the personal questions on the page of a woman's magazine to discover if we are good social mixers, we may, nevertheless, rebel when the income tax return is pushed through our letter-boxes.

The use of statistics to persuade people to certain opinions is a popular practice. The danger lies in the fact that there are good and bad statistics and statisticians, just as there are good and bad doctors. The saying that 'figures cannot lie' contrasts well with the cynical comment by Disraeli that, 'There are lies, damned lies, and statistics'. Neither of these statements is necessarily true. Yet we know that:

1. Statistics may be collected carelessly.
2. Figures may be gathered for one purpose and used (by someone else) for a different, probably contradictory, purpose.
3. Statistics may be processed wrongly.
4. Statistics may be interpreted wrongly.

It is necessary, although only remotely possible, in a successful democracy for the person in the street to be able to detect lies in figures as well as in words. To do this, he or she should know how much notice to pay to the statisticians' figures, and also how to deal with their flattering requests for a personal opinion and the details of daily life. It is more necessary for students to be able to detect such fallacies, because as well as taking an examination in statistics, they might well be joining the growing mass of questioners, enquirers and interrogators who descend on the public, absorb masses of details, process them and finally disgorge them on a public which is becoming increasingly suspicious of matters presented to it in print, on radio or on television.

2

Collection of data

Primary data consist of figures collected at first hand (by the methods described later in this chapter) in order to satisfy the purposes of a particular statistical enquiry. Examples in the government field are the Census of Population and the Index of Retail Prices. In the field of business, market research enquiries are the most obvious example.

Primary and secondary data

Secondary data consist of figures which were collected originally to satisfy a particular enquiry, but have been used now, at second-hand, as the basis for a different enquiry (usually because the cost of collecting new data would have been too great, and the second enquirer believes that the original figures will adequately suit the need).

Examples in the government field are figures which go to make up the Balance of Payments calculation, where existing figures are collected for this purpose from many sources, principally from the declarations to customs officials by exporters and importers. In the field of business, an example would be figures relating to productivity (compiled from cost accountants' records and sales records), but this is only one use of secondary data in a field where many departments rely on each other to supply figures for individual purposes.

Note. Secondary data must not be confused with secondary statistics (see page 99).

In this section we are concerned mainly with the methods used to collect primary data. The sources and methods used in dealing with secondary data are dealt with in Chapter 21.

Methods of the survey

The various stages of a survey may be classified as follows:

1. Purpose of the survey.
2. Sampling methods.
3. Methods of collection.
4. Pilot surveys.

Purpose of the survey

At the very outset it is essential to consider the purpose of the survey because this will not only affect the questions we ask, but it will also dictate the methods to be used.

The sections that follow after this stage are mainly mechanical in a sense, but the cost of the whole operation in time and money depends on detailed, careful and thorough planning at the beginning. Even if the scope and nature of the survey is decided, it must be realized that, in either business or government circles, it is usual to work to a target date and within a limited budget, and initial planning may suggest a point at which a halt must be called to further enquiries. It is often a great temptation to extend the scope of questioning, for example, and to add subsidiary lines of enquiry to the main line on the assumption that a slightly larger questionnaire will do no great harm. Possibly this might be the case (see the questionnaire, page 11). But it is of great importance that all such problems and possibilities be explored at the beginning. Once decided, the terms of reference must be rigidly followed.

Sampling methods

Surveys may be carried out on people or on things. On the production line we could apply a survey technique to control quality; in society we may wish to survey the expenditure habits of the population.

Statistical techniques have improved so greatly that it is not now necessary to investigate all the items in any group from which we may wish to collect information. Instead, we can carry out a survey of a population by merely investigating a fraction, i.e. a sample, of that population. The whole of the group under investigation is known as the *population*, even though the group may consist of factories in north west England, or the people in the nation itself. The small number of items from this population actually examined is called the *sample*.

It is not intended here to describe the mathematical theory of sampling, but simply to say that we can tell, with a high probability of being correct, what the characteristics of a population are by investigating only a sample of that population, provided that we observe certain rules. These are:

1. Our sample must be of *at least a certain size*. Generally, the larger the sample the more reliance we can place on our results as being a true cross-section of the population.
2. We must choose the sample from the population in such a way that *each member of the population has an equal chance of being chosen*; this is known as random sampling.

Provided that our sample is truly random (and, of course, large enough) we will get a representative cross-section. This is really much harder to achieve than it sounds.

For example, if we were to try to choose a random sample of the people walking down the main street, in order to ask them questions on their leisure habits, we should probably find that after a time we tended to choose the slower walkers, the older people or the people with time to spare and chat. If we questioned a sufficient number of people and analysed the results, we might conclude that the way most people in the area spent their leisure time was in playing bingo, walking round the park and watching television. We would tend to miss the busy adult males and, if we took our sample in mid-morning on a weekday, we would miss all the young people. Thus, a great range of leisure activities from ballroom dancing to pigeon racing would find little place in our survey. Our results would be biased in favour of certain types of people, i.e. they would not be truly representative of the whole. In other words, it would not be a random sample, because some people (business executives, factory workers, school children, etc.) would have little or no chance of being selected.

Had we conducted our survey on a certain Saturday afternoon, hoping to catch such people as we missed before, we might conclude from our results that people in the area were 'sport mad'. Preliminary planning might have warned us not to stand on the road which led from the bus stop to the football ground on that particular Saturday.

This, of course, would have been a glaring mistake. But the techniques of sampling, which enable us to obtain the maximum of information with the minimum of effort, impose rather stricter rules than this on the investigator. Probably it is impossible to make a perfectly random selection. Yet statisticians are constantly trying to achieve the impossible.

One of the methods they have evolved is that of *lottery sampling*. By this method, if one is selecting a sample of people, one should write down the name of each member of the population on a piece of paper, place the papers in a revolving lottery drum, and then pick as many as the sample requires from the drum. Sometimes the *constant skip* (or *systematic*) *method* of selection may be used. This is the selection of a sample by choosing, say, every tenth name from a voting list, or every seventeenth house from a street plan.

This question of better random selection methods and the avoidance of bias is the subject of extremely careful study among statisticians. One further example of the introduction of bias might be given.

It is intended to investigate local tradesmen in an area to discover whether they were born in the area or whether they moved in from elsewhere. To select the names for a sample of 1000, we might use a copy of the classified trades telephone directory, simply open it 1000 times at random, and pick a name each time. This would be our list, and we could probably ring up on the spot to ask our questions.

This method would *not* give us a random sample. Bias might be introduced in the following ways:

1. Tradespeople who are not on the telephone, e.g. small shopkeepers, would have no chance of selection.
2. The telephone directory, through constant use, falls open at some pages more readily than at others. Therefore, traders' names on rarely opened pages, e.g. violin makers, would have a smaller chance of selection.
3. Our eye might be attracted, subconsciously, to the more unusual names, e.g. foreigners, on the pages. Thus, anyone with a common name would stand less chance of selection.

7

4. Some tradesmen who have a telephone might not be in the directory, e.g. newcomers within the current year.
5. The directory might be out of date.
6. Some traders might be entered more than once, e.g. cross-referenced under a split trade.

There might also be other sources of bias besides the ones on the above list.

From most of these comments a certain conclusion can be drawn; that is, that the human being is a very poor instrument for the selection of a random sample. In fact, most errors of bias tend to occur because of the human element.

Different methods of applying random sampling may be used, depending on the kind of population we wish to survey.

STRATIFIED SAMPLING

People fall naturally into different groups according to their social background, sex, age, etc. It is obvious that a survey, say, on pop music, might well draw different opinions and answers from each different group, i.e. not merely from different individuals. For example, old people may not appreciate pop music as much as the young.

We might conduct a survey in a town where the population is largely composed of old people, e.g. 30 out of every 100 are over 60 years old, and 10 out of every 100 are under 21. To stratify our sample in respect of, say, age (i.e. divide the population into age strata) we should have to apportion our interviews as follows in a sample of 1000:

- 300 interviews of people over 60 years;
- 600 interviews of people between 21 and 60 years;
- 100 interviews of people under 21 years.

This would give us a stratified sample if we chose 300 people *at random* from that part of the population over 60 years, 600 from that part of the population between 21 and 60 years, etc.

Each group would then be better represented (in the correct proportion in the sample) than if we had merely chosen 1000 people from the whole population. In addition to this, it might be a subsidiary part of the purpose of our survey to compare the answers of the particular groups.

MULTISTAGE SAMPLING

This is another way of using random choice to select a sample from a very large or widespread population. It can be illustrated as follows.

From 50 states in the USA, we might choose a random sample of five states. Then from each of these five states, we would choose a random sample of 20 towns. Then from each of these 20 towns, we would choose a random sample of 100 people. Thus, we should have 10 000 interviews. Provided that we kept the sample large enough to cover the large entire population, we should be able to concentrate our interviews with a minimum expenditure of time and cost. The actual selection of the sample would also be much easier.

This method is almost the only practical one when we wish to make a survey of a large population on a limited budget. There are several variations of the method of multi-stage sampling, the most popular being to relate the number of interviews in particular areas proportionately to the population in those areas. In this way, we can reduce the bias likely to creep in because the many small urban communities have a better chance of being selected than the few large urban communities.

QUOTA SAMPLING

In this case, the interviewer is not given a list of the selected names in the sample, but is asked to interview a number (quota) of people (at his own discretion) who fall into certain categories:

Table 2.1

Column 1	Column 2	Column 3	Column 4
Category	Types to be selected	Quota to be interviewed	Totals
Possessing car (value over £10 000)	Directors	10	
Telephone	Managers	10	30
Deep freeze	Professional men	10	
Detached house (owner)			
Possessing car (value £10 000 or less)	Teachers	60	
	Civil Servants	60	
Television	Shopkeepers	60	240
Semi-detached house (owner)	Skilled artisans	60	
No car	Transport workers	100	
Television	Shop assistants	130	730
Rented house	Labourers	500	
		TOTAL	1000

In a way, this form is similar to stratified sampling, the main difference being that the interviewer can choose the individuals to be interviewed. Here lies an important source of bias as human random selection is extremely faulty. In particular, the interviewers may interpret the types differently. For example, a turf accountant is not usually regarded as a professional, but a chartered accountant is. This difficulty may be overcome by providing a list of definitions of each type which the interviewer must follow. Another fault is that the interviewer, to save time, might choose a large proportion of any category from one particular area. To overcome this weakness the quota can be stratified so that the interviewer may pick only certain types in each category (see column 2).

Methods of collection   There are two basic methods of collecting data—the personal interview and the questionnaire.

THE PERSONAL INTERVIEW

A disadvantage in the method of collecting data by means of the personal interview is the high cost, not merely in the expenses of wages for staff, transport and allowances, but also in the cost of any staff training which may be given.

Most of the points raised in the next section (the questionnaire) apply equally to the personal interview, but the most important difference lies in the necessary personal qualities an interviewer should possess. The qualities most desired are listed here.

(a) *Skill*

The interviewer is usually the worst cause of bias in a survey, so it is important to provide an initial short course of training in order to avoid the most serious pitfalls. A modern trend in interviewing which would require longer and more expensive training is 'depth-interviewing'. The tendency here is towards longer and deeper questioning, which probes the hidden factors that lead people to answer questions in a particular way.

(b) *Character*

Various qualities of character are desirable in an interviewer, and these may be listed briefly as follows:

(i) *Tact* It is important neither to antagonize nor to flatter interviewees because they may tend to depart from the truth under emotional influence or stress. Maintaining a calm, unflustered, and enquiring attitude at the end of a long hard day is difficult to do when one is repeating a set of questions for the fiftieth time. Nevertheless, the interview should be restricted to a rational and thoughtful imparting of information, and it is the interviewer's responsibility to maintain this state of affairs.

(ii) *Accuracy* Once the list of interviewees has been drawn up by the central office, the interviewer should keep rigidly to the list. An interviewer who omits to interview a selected person because that person lives at the end of a particularly long garden path or because a growling rottweiler guards the entrance, will undoubtedly introduce bias into the carefully prepared list.

Accuracy also means arithmetical accuracy in any short calculations which the interviewer is asked to make, as well as accuracy in recording the answers given and the checking of lists.

(iii) *Amiability* Many people do not respond well to proposals to investigate their private lives and habits. This imposes a special task on the interviewer— that of having a pleasing and sociable personality without distorting the interview and the answers by being excessively charming or effusive.

(iv) *Neutrality* It is obvious, by now, that a very special kind of person is required for this job (some might say a genius or a saint). Needless to say, the demand for such people exceeds the supply (at the wages offered), and so great a degree of bias is believed to be introduced by inadequate interviewing, that several studies have been carried out on errors of bias due to interviewers.

When we think of the possible bias in matters of colour, sex, race, religion and politics which may be subconsciously introduced by the interviewer into the interview, the dangers here are apparent. Should interviewers have strong convictions? Is the present general policy of employing part-time interviewers a good one? We might well wonder how impassively neutral persons can be tactful and amiable in all situations. It may be comforting to know that, as in most jobs, the ideal is unattainable.

There are obvious drawbacks to the task of making personal contact. For example, evening callers are likely to find a high proportion of stay-at-homes, Saturday afternoon callers will tend to miss the sports fans, etc. In such cases, bias will result, unless the timing of the calls is carefully recorded, and further calls are made on the absentees. Bank holiday and Christmas interview visits are likely to reap their own peculiar rewards!

While the questionnaire method does not suffer from these disadvantages, it is much less likely to produce as high a response as direct personal tracking down, and the confrontation method of the personal interview.

THE QUESTIONNAIRE

Let us consider here the questionnaire delivered by post.

Advantages

This method of enquiry is, of course, much less costly to operate than the personal interview method. This means that we can increase the size of our population (i.e. extend the area of our survey) or we can take a larger sample from a given population. By making our sample larger we stand a chance of making our results more reliable, i.e. more representative.

This method, above all, is free from any personal bias which might be introduced by an interviewer.

Disadvantages

We cannot count on a high proportion of replies, unless there is a legal obligation to reply (as in the Census of Population) or an inducement, usually goods or cash, is offered (a cash contribution was offered to each spender in households which cooperated in the Family Expenditure Surveys). A response of 20 per cent is considered quite good in the average survey. This lack of response may lead to bias, called 'non-response bias'; i.e. our results may not include a certain type of person with whom we wish to make contact and from whom we want information. This may be the type of person who usually ignores circulars. Often the kind of people who *do* reply are of certain types, e.g. people who are too timid to resist, people who enjoy filling in forms, people with a strong sense of public duty (often the opening appeal of questionnaires asks for 'your cooperation', etc.). As long as certain types are included in the replies, and certain types are excluded, we will not get a representative sample. A further

11

disadvantage is that no interviewer is on hand to explain the questionnaire, or to ensure that it is filled in completely or correctly.

Features of a good questionnaire

The popularity of this type of enquiry, which is often included nowadays in cut-out pages of magazines and newspapers, has led to much investigation into the extremely difficult task of achieving better design and question-phrasing. A list of 'do's and 'don'ts' can be made, as follows:

1. Keep the questions themselves, and their number, short.
2. Phrase the questions and the instructions, if any, as unambiguously as possible. An example of a badly phrased question would be: 'Do you make a habit of drinking?'
3. Do not use unusual, pompous, or technical words.
   Wrong: 'Indicate your marital status.'
   It is a good plan to aim at the lowest intelligence and educational level of the population which you are investigating, without making this too obvious.
4. Do not use leading questions, or words which are currently used in an emotional or abusive way.
   Wrong: 'Are you in favour of even more privatization?'
5. Include check questions, where you expect some people to give unreliable answers. For example, people often have a habit of rounding off ages to the lowest 10 below (see Cumulative Errors, page 96). In this case, the age (in years and months) can be asked for near the beginning of the questionnaire, and the check question—'Please give your date of birth'—can be inserted later in the form.
6. Always state the precise units in which you require the answer, otherwise the results, when you come to tabulate them, will be useless. But do not ask irritating questions that involve an unnecessarily high degree of accuracy.
   Wrong: 'How much gas did you consume last month?'
   Does this mean therms, money's worth? 'Consume' means 'eat' to the average person!
7. Do not ask questions which rely too much on memory. Accurate memory of past events fades much more quickly than people realize or are willing to admit.
   Wrong: 'How many days were you absent from work because of sickness in the year ending 31 December last?'
8. Try to ask simple questions which can be answered by a yes, a no or a figure.
9. Do not ask the average man or woman to write an essay for you. This type of question usually indicates lazy thinking on the part of the compiler.
   Wrong: 'Please state, as fully as you can, the reasons why you prefer to use biological soap powder'.
10. 'Cafeteria' questions (questions in which the respondent chooses his reply from a range of alternatives) should be of the right kind:

Right

Place a tick against the kind of home in which you live:

(a) flat (in a block of flats).
(b) bungalow.
(c) semi-detached house.
(d) terraced house.
(e) detached house.

Note. It is assumed that the survey includes only these types of dwelling.

Wrong

State whether you think that unemployment will:

(a) rise.
(b) fall.
(c) stay about the same during the next six months.

The second cafeteria question is wrong because:

(i) Most people, with little knowledge of what is a rather technical question, will answer (*c*). They will take the middle course rather than betray their ignorance. In any case, there is no 'don't know' alternative listed.
(ii) 'Rise' and 'fall', in this context, mean virtually nothing. For example, does a rise mean an increase of one per cent or 300 per cent? Any calculations made from the answers to such questions will also mean virtually nothing.

11. Do not ask questions which call for calculations.
   Wrong: 'How much per annum do you spend on smoking?'
   Even if a person knows roughly how much he or she spends per week, the result of multiplying this by 52 may be inaccurate. The respondent may also jib at the task.

There are many more points to watch for when compiling and designing a questionnaire. Some are obvious. For example:

Date of birth ....

Give, with dates, names
and addresses, all the
schools you have attended ....

Others are not so obvious.

Remember that you are asking the public to spend time and effort for no apparent reward. You must, therefore, make the questionnaire attractive and

13

sensible. Explain the purpose of the questionnaire, arrange the questions in a logical sequence, if possible, promise to keep the results confidential, and either collect it or pay the postage for its return.

Pilot surveys

Pilot surveys are carried out before the actual survey itself, although sometimes the limited budget and the time available will not be sufficient for the added cost of a pilot survey. Also, if a similar kind of survey has been carried out before, a pilot survey might not show anything useful. If the actual survey is on a big scale, and is likely to be an expensive one, there is much to be said for conducting a pilot survey in the early stages.

The pilot survey is essentially a small-scale replica of the actual survey which, except for coverage, it should duplicate as nearly as possible. Its use does not lie in the actual replies it brings, but in the lessons it teaches. It may reveal faults and weaknesses in proposed methods, especially as it is usually carried out by a highly trained permanent staff. The pilot survey might amply repay the extra time and money spent on it, because it would be far more expensive to have to abandon the actual survey, once begun, if major faults have gone unnoticed. What may seem perfectly clear to officials in the survey office, may appear vague and ambiguous when presented to the person in the street. The only sure check is to test it under actual conditions.

Computers and data collection

Telephone surveys

Computers now play an enormous role in the analysis of data once collected, as we shall see in future sections and chapters. They can also play a part in the actual collection of data. For instance, some large market research organizations, especially in the USA, use computers in the selection of households to be included in telephone surveys. Surveying households by telephone interview gives some of the benefits of the traditional face-to-face interview while drastically reducing the cost. There are also disadvantages, e.g. less likelihood of rapport being established between interviewer and respondent. In countries where nearly all households have a telephone, or in surveys where the target population to be interviewed excludes households or businesses without telephones, the threat of bias in sample design is smaller (see section on Sampling methods on page 213).

The computer can be used to generate telephone numbers at random. Behind many computer games there is an element of randomness, and this familiar feature of computers in having a random number generator (or, more exactly, a pseudo-random number times generator) can be used in the survey context. The computer can be used to select, and even dial, the next potential respondent. Because we are generating numbers to be dialled from within the computer, rather than using the conventional paper telephone directory, some of the problems of sample design mentioned on pages 6 and 8 are avoided—e.g. new telephone customers have an equal chance of selection, as do households with ex-directory numbers. Of course, some problems remain (e.g. 'households' which share a single telephone—perhaps bed-sit tenants in a boarding house), and such problems give some threat of bias.

The telephone survey also avoids the familiar 'paper and pencil' interview format in which the interviewer records the respondent's answers on paper, and replaces it with more direct entry of the answers on a *visual display unit* (*VDU*) computer screen. Most of us are already familiar with commercial electronic 'forms' displayed on VDUs in travel agencies, on which a counter clerk enters our reservations for train, coach or holiday. The telephone interviewer can also utilize such a form, thereby omitting one stage of data manipulation (i.e. the keying of data into the computer from the original paper record) and removing a frequent source of potential error—the misreading and miskeying of the data from paper records.

Data to be analysed statistically can also be gathered on computer by direct data 'capture'. Supermarket checkouts which use barcodes on goods passed across a barcode reader in order to produce the customer's bill, represent a now common experience for shoppers. The data gathered can be used not only for automatic restocking and reordering, but also to analyse the 'performance' of individual product lines and retail outlets, and even of individual checkout workers (e.g. the average number of goods or customers dealt with per hour). **Direct data capture**

Direct data capture methods can sometimes be used to replace or complement traditional paper and pencil surveys. Broadcasters often rely upon the 'morning after' quota sample survey (see page 9) and the smaller 'panel' of volunteer diarists to assess who saw/heard which programmes, and what respondents thought of them. A newer possibility is to have a small computer automatically registering which channel the television in a volunteer household is tuned to, at what time, and for how long. If the household's computer is connected to a telephone line, the main computer at the research headquarters can periodically record a general database of viewing habits in all volunteer households. All that the household volunteers have to do is record, perhaps every 30 minutes, how many people are viewing the set (with their ages, sex and other relevant information). The main part of the data collection is done automatically and electronically.

One of the advantages of this system is that it requires less thought and effort by volunteers than does the full diary method. This means that there is less likelihood of household viewing being unrecorded. Because the automatic data capture is less intrusive than the requirement to complete a viewing diary, there is perhaps less danger that households will change their viewing habits to make themselves appear more 'cultured' or somehow more 'typical'. Of course, some problems remain—e.g. the bias inherent in using small samples of self-selected volunteers, the problems of household members viewing programmes outside the computer-connected set, and the problem of 'viewers' who are really concentrating on other activities, like teatime!

OPTICAL MARK RECOGNITION

Some older methods of computer-oriented data collection are still in use. The method of inputting data into a computer known as *optical mark recognition* (OMR) requires the use of a special paper form. The respondent uses a pencil to mark a small box which corresponds to the answer category selected— perhaps one of four numbered boxes which represent the responses offered to a question in a multiple-choice test. A form may contain tens of such banks of boxes—for as many questions—plus spaces where the respondent or candidate can record name or identifying number. The forms of all respondents are gathered together and fed through an 'optical mark reader'. This reader works by aiming a thin beam of light at the paper form and registering the reflected light. Where a pencil mark, corresponding to the answer selected, prevents light being reflected from one box, the computer program understands that this was the respondent's choice. It is therefore the position of the mark on the paper, preventing the reflection of light, which is crucial in an optical mark reader.

This method of data recording and input has been common in multiple-choice tests, particularly in schools and colleges in the USA. The British Open University conducts computer-marked assignments using this technique. An EDPAC answer sheet is reproduced opposite as an example of an OMR form.

Optical mark recognition can be used in other administrative contexts. College enrolments have been made with OMR forms. The advantage is that information is supplied in a 'machine-readable' form, thus allowing computer processing (e.g. data passed automatically to a timetabling program, or to a program that will tabulate the previous qualifications and backgrounds of applicants) without the slow and expensive need for data to be keyed into the computer from conventional application forms by keyboard operators. However, one disadvantage is that the OMR forms appear strange, and at first sight intimidating. If the respondents are not already used to OMR techniques it is best to have advisers and checkers present to help those answering the questions.

Surveys by computer

By the 1990s, both the social role of 'survey respondent' and the use of VDU/keyboard computer interaction are well known to citizens of industrial countries. Most of us are used to providing data direct to banks (in order to obtain a service) by keying in requests ourselves to a 'hole-in-the-wall' cash dispenser.

Increasingly, surveys will be conducted in which respondents enter information directly to the computer, reading the question from the VDU and entering their reply via a conventional keyboard, or by pressing a touch-sensitive screen (here the screen becomes both input and output device for the computer). Compact disk storage for computers (of data, programs, speech), plus voice recognition computers have already been developed, and we can easily foresee some surveys being conducted in which respondents listen to and speak to computers, saying rather than typing simple responses like 'yes', 'no', or specific numbers such as 'three' or 'five'.

Since the mid-1980s, it has been possible for researchers to purchase computer programs which simplify some of the major stages of survey enquiry. Some software packages for micro-computers (e.g. **Courseview, Sphinx**) aid the researcher at three stages—question formulation, data gathering, and data analysis and report writing. These packages help researchers conduct a survey. At the first stage, the Courseview package proposes a series of preformulated

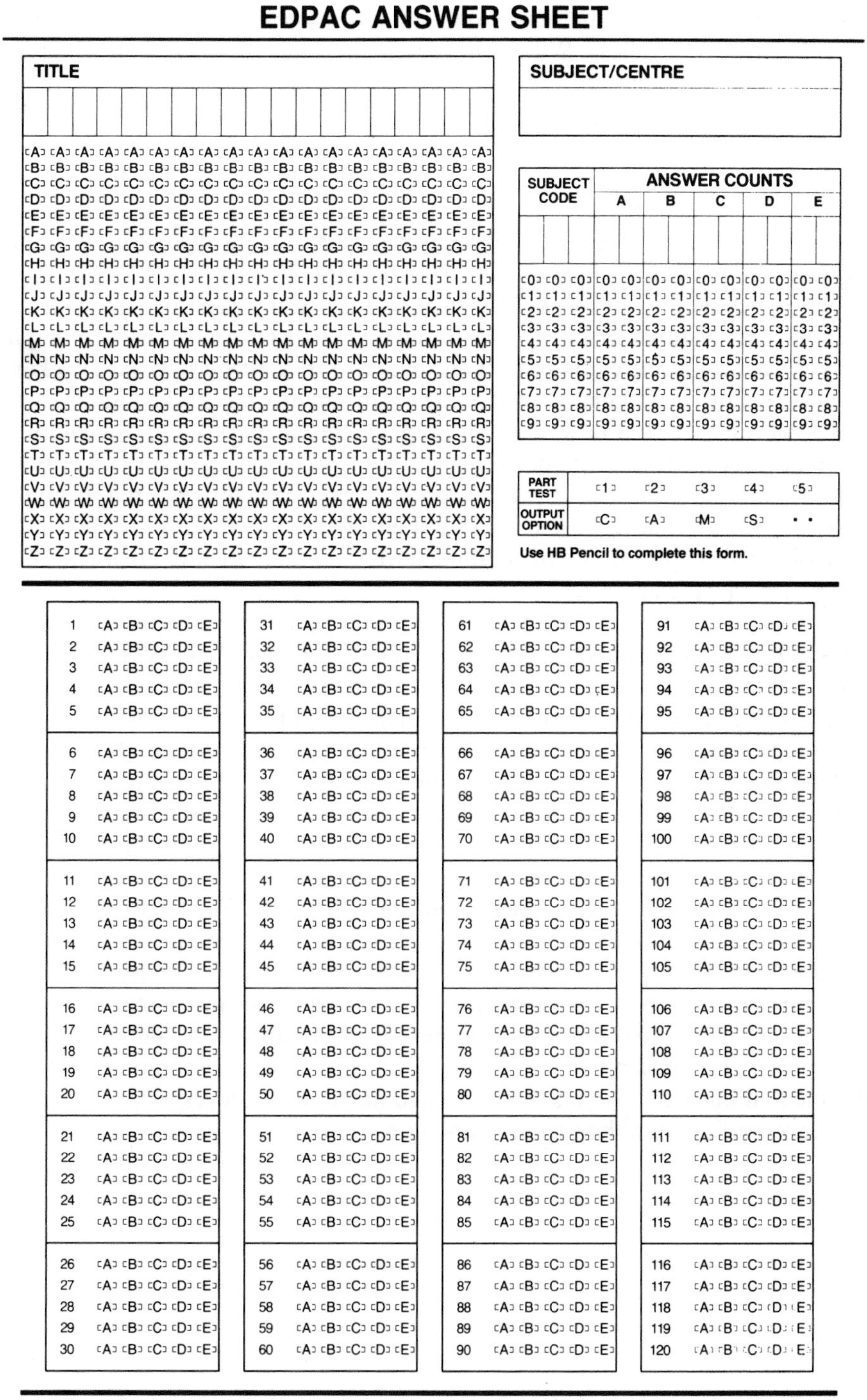

EDPAC ANSWER SHEET

TITLE

SUBJECT/CENTRE

SUBJECT CODE	ANSWER COUNTS				
	A	B	C	D	E

PART TEST	□1□	□2□	□3□	□4□	□5□
OUTPUT OPTION	□C□	□A□	□M□	□S□	· ·

Use HB Pencil to complete this form.

Fig. 2.1 EDPAC answer sheet

questions on various subthemes, allowing the researcher to delete questions, or to add some new ones. When the researcher has finished compiling the questionnaire from the mix of pre-written and new questions, the questionnaire is left on the computer for the survey respondents to answer. At this second stage the respondents supply answers (via VDU and keyboard) at their leisure. Answers are stored, and the program records how many respondents have completed the questionnaire. When the researcher is satisfied that a large enough sample of respondents has replied, there is a third stage in which the package automatically analyses the data. Tables are produced which group replies by conventional variables, such as age and sex, thus allowing the researcher to distinguish between different types of customer. More complex analysis can be conducted by passing the data to a specialized statistical package.

More facilities are available to researchers to simplify and make less costly the process of research. It is now usual for academic researchers to deposit computer magnetic tapes containing social survey data with the appropriate national archive. In the UK, this is the Economic and Social Research Council's (ESRC's) Data Archive based at Essex University. These data are then made available (with appropriate ethical safeguards) to other researchers for 'secondary analysis' (see page 99).

It is also possible to buy 'ready-made' lists of relevant groups on whom surveys (or calls for charity donations, etc.) can be made. This practice is common in the USA, and is a sign of how much of our lives has come to be recorded electronically. For instance, a Chicago publishing company may be thinking of launching a new society magazine. It wishes to conduct some market research to find the present reading habits of its likely readership—the wealthy of Chicago. Is there room in the market for a new magazine? What would potential readers want in a new publication? It is possible to buy information, from commercial data agencies, on the names and addresses of all those in the Chicago area who have a credit rating above so many thousands of dollars. We might also specify that we want details only on those who are registered Republican voters (data available thanks to the USA voter registration system). These rich trails disclosing people's lives are available in the first place because of the computer. Further, our ability to select special subsets (by credit rating, residence and voter registration) is made possible thanks to the power of the computer. Having obtained all our rich, Republican, Chicago names and addresses we can take a sample from this 'sampling frame', and conduct our survey.

3

Computers and software packages

This present chapter is a general introduction to the computer as used for statistical analysis. The next chapter, Chapter 4, deals specifically with preparing data for analysis by computer.

Computer hardware

The computer is now a commonplace general-purpose machine. It is beyond the scope of this book to describe how a computer works, or all the forms of computer equipment or *hardware*. Basically, the computer has four different types of component—*input devices*, *output devices*, a *central processing unit* and *storage devices* (for programs and data). We shall assume that readers have met, if only in a passive way, a computer for which commands and data can be input from a keyboard. The results of data manipulation by the central processing unit of the computer (the part that does all the calculation!) are output to a familiar *visual display unit* (*VDU*) screen, and often to paper ('hard copy') by a printer or more specialized *plotter*. The VDU also aids input by reproducing on the screen everything typed from the keyboard. Data and programs can be stored on various media, including *magnetic tape* and *magnetic disk*. Some storage devices are built into the computer, e.g. *hard disks*. *Back-up storage* on small portable *diskettes* is now familiar to most people in education and business (see below for diskette care).

MAINFRAMES AND MICROS

Business and education computer users may meet computers in various forms. The *mainframe* computer is a powerful machine, with huge computational and storage facilities. It is kept and serviced in a specially protected environment. It can serve many different users simultaneously, each user possibly employing a different application (e.g. wordprocessing, statistical analysis, computer-aided design (CAD)). The keyboard/VDU combination in front of the user is called a *terminal*. The terminal may be 'dumb'—i.e. have no computing power of its own unless linked to the mainframe kept in another room or building. The mainframe may have scores of terminals and printers, plus other input and output devices, attached to it. The *mini* computer is a smaller version of the mainframe.

The *micro-computer* is the desk-top machine that is essentially self-contained —it has its own complete set of input, output and storage devices, plus a central processor. The micro may sometimes act as an '*intelligent terminal*', i.e. it may act as a terminal to a mainframe or mini, while retaining its own computing power. Micro-computers may also be similarly *networked* to one another.

Computers do not all use the same *operating system* (language of command). Micro-computers for education or business are likely to be **Apple Macintosh** or **IBM PC** machines, with their associated distinct operating systems. The PC ('personal computer') machine created by the IBM firm has now been copied by many other manufacturers. The term 'PC' is therefore now a general label for a whole type of computer. Among micro-computers, this book concentrates on PC-style computers rather than other types.

Our statistical analyses in this book will be made with two related types of program—the *statistical package* and the *spreadsheet*. A statistical package is a large program, or suite of programs, dedicated to statistical calculation. A spreadsheet is a program used for more general mathematical calculation (e.g. accounts), including statistics.

The statistical package and spreadsheet can be thought of as powerful calculators. Whereas we use the familiar everyday handheld calculator by pressing individual keys to call arithmetic and statistical functions (e.g. the percentage key), with the computer statistical package or spreadsheet we have to key in more complex commands from the keyboard.

Statistical packages, originally written for mainframes, are now commonly available for the PC-style computer. Statistical packages like **SPSS/PC+** are adaptations of programs first written before the micro-computer was developed. Programs are often given the slang name *software* (as opposed to the hardware physical components of the computer). Large statistical software like **MINITAB** is often called a *package* because it is more exactly a bundle of interrelated programs than a single program.

STORAGE FOR MICROS

Users of micro-computers can save programs and data on *diskettes*, normally in one of two sizes—5.25 inch or 3.5 inch. The smaller diskettes are more advanced and can actually store more information than the larger ones. The 5.25 inch diskettes, in particular, are often known as 'floppy disks'. The computer can read information from, or store information on, diskettes when they are placed in the *disk drive*, rather like a compact disk reader mechanism. Diskettes should be handled with care. They are easily damaged and rendered unusable, especially the 5.25 inch 'floppies' which have no protective hard plastic sleeve and cover plate.

Before diskettes can be made to store information they first have to be *formatted*, i.e. electronically prepared for the storage conventions of your particular type of computer. This is done by calling up the FORMAT program and placing the appropriate diskette in the disk drive. The exact procedure will vary between computer installations.

You will find that the computing sections in this book are very practically oriented. There are lots of worked examples. As far as possible (depending on which packages or programs you have available), you should try the worked examples for yourself. Once you have obtained the same answer as we have, you can go on with confidence to new examples.

Conventions

Computing instructions in this book take a standard form. Where special keys—as opposed to the normal letter and digit (alphanumeric) keys—are to be pressed, they appear in **bold** type within the ⟨ ⟩ symbols. Hence, the key which carries the label 'Esc' will appear in the text as ⟨**Esc**⟩. The key labelled 'F1' will appear as ⟨**F1**⟩ when we want to indicate that it should be pressed in a sequence of commands. Commands to the computer, other than the special keys like ⟨**Esc**⟩ and ⟨**F1**⟩, are normally followed by pressing the 'Return' key (⟨**RETURN**⟩, ⟨**Enter**⟩ or ⟨**↵**⟩). Since it is tedious to repeat this instruction continually, it is largely omitted in the text. In some packages, e.g. the spreadsheet **Lotus 1–2–3**, some ordinary symbol keys like the forward slash or oblique ('/') are important in evoking commands to the computer. Where such keys are used as command keys rather than as ordinary symbols, they too appear within ⟨ ⟩ brackets, e.g. ⟨**/**⟩.

Commands to be typed within statistical packages are usually indicated in this text by *upper case* (capital) letters. Commands to the operating system of the computer, outside the package, are usually given in *lower case*. This purely printing convention is used simply for clarity—to indicate what we are doing. In computing practice, upper and lower case may be used interchangeably in all the applications discussed in this book. As far as possible, output from computer processing is reproduced as it appears on the screen.

Now that we are reasonably familiar with computer hardware, we turn our attention to the software—programs and the complex programs known as 'packages'. Using the computer for statistical exploration normally involves having the data analysed by a package. The software is likely to be a dedicated statistical package or it may be a spreadsheet.

Programs and packages

There is nothing mysterious about the idea of a computer program or software package. A program is nothing more than a set of instructions written in a language that the computer will understand. A program is like a knitting pattern, a cooking recipe or a car maintenance manual—just instructions. Like these last three examples, particularly the knitting pattern, the language that is employed to activate the computer has conventions and a grammar of its own.

The software packages that we deal with in this book can do nothing without the input of data to work on—just as the knitting pattern is useless without a ball of wool. The next chapter outlines how to get the data successfully into a stored computer data file. Sometimes we may wish to key in the data at the same time as the commands to the statistical package. This too is possible (see Chapter 5). In any case, one way or another, the data has to be supplied by the user. The software package is an independent tool, but what it is asked to work on is at the user's discretion.

As we shall see, there are many packages available, and there are many types of mainframe and micro-computer. This book illustrates the use of various mainframe and micro versions of packages. It cannot illustrate the use of every package on every type of computer. The examples supplied in the book have been tested and found to work, as reproduced in these pages, on the computers and packages available to the authors. Undoubtedly, some readers will find that, on the computers they use, some of the packages are accessed differently or even work differently. The data handling workings of the mainframe **SPSS**ˣ package in particular are likely to vary between computer installations. So, you may need further information specific to your computer installation.

Where possible we have tried to illustrate the use of both mainframe and micro versions of packages. Mainframe and PC versions of the **MINITAB** package are very similar except for the improved data entry facility of the PC version. The PC version of SPSSˣ is **SPSS/PC+**. These two versions are less alike.

INTERACTIVE VERSUS NON-INTERACTIVE

Computers have two basic ways of working—*interactive* and *batch* (non-interactive). The basic style of the MINITAB statistical package is interactive. The basic style of the SPSSˣ statistical package is batch or non-interactive.

With interactive working, the computer processes the commands from the keyboard one at a time, every time that the ⟨**RETURN**⟩ key is pressed. An interactive package like MINITAB can be thought of as a clothes shop where the customer goes in and tries on a few possibilities. Mistakes are immediately obvious (in the mirror or VDU screen). We simply try again until we get the commands, or the choice of clothes, correct.

With batch working, the commands are entered into the computer and temporarily stored before being sent off to be acted upon as a complete parcel. Results may not be sent back automatically to the screen, but may be stored as an *output file* within the computer. This file can later be called to the screen and/or sent to a paper printer. A non-interactive package is more like mail-order—the transaction is more obviously formal, 'bureaucratic' and apparently removed from instant trial and error. Typically with SPSSˣ, like a mail-order purchase, our 'order' (the commands to produce the needed statistics) is drawn up in advance. The order or *job* is then sent to the package or mail-order company. The goods or statistical results are sent back. We open the return parcel or output file in which they arrive. If there has been a mistake we have to amend our order form or 'job file' and resubmit it. The revised order is processed and a second return parcel or output file of results dispatched to us. Again, we open up the package, and either accept the contents or else revise our order and submit it again.

SPREADSHEETS AND STATISTICS

The *spreadsheet* package or program is now a familiar part of desktop comput-ing. Typical spreadsheets are **Lotus 1-2-3** and **AS-Easy-AS**. Sometimes the spreadsheet is part of an even larger package of office software known as an *integrated package*. An integrated package usually has a wordprocessing facility, a database (an electronic filing cabinet) and a spreadsheet program. **Smart** and **Works** are examples of integrated packages.

A spreadsheet is generally less specialized than the statistical package, allowing the user to manipulate with ease many sorts of quantitative data—financial, mathematical and statistical. The user enters the data into the *worksheet*, which is a table of *cells* in rows and columns. Above or below the worksheet is an area of the screen (*command area*) in which the user can call up or type a wide variety of commands to define, store, manipulate or print the data and the statistical manipulations. A *cursor*, a small flashing bar or highlight on the screen, indicates the user's position on the worksheet. The worksheet mimics the way in which the human worker with pencil and paper would draw up the data into a table with rows and columns. However, whereas the worker with pencil and paper may be equipped with a small electronic calculator, the spreadsheet user obtains statistics from the package by typing in his or her own formulae, or by calling up pre-written and stored formulae permanently installed in the package.

A spreadsheet worksheet with data, labels and results, as these elements would appear on the screen, is given in Fig. 3.1. Spreadsheets share very similar screen displays.

	.......A/	......B/	......C/	......D/	.......E/	......F
1	FIRM	Employees	Profit92	Profit93	Both92+93	Per Employee
2						
3	SFB plc	12	23000	18900	41900	3491.666
4	Teeswork	45	1200	23000	24200	537.7777
5	Autocan	108	−25000	−45078	−70078	−648.870
6	NHFG	467	−135900	678000	542100	1160.813
7	Cleveway	25	140000	−2300	137700	5508
8	Easterside	67	−36790	1500	−35290	−526.716
9						
10	TOTAL	724	−33490			
11	MEAN	120.6666	−5581.66			

Fig. 3.1 Spreadsheet example worksheet (screen display)

The spreadsheet's style is similar to a dedicated interactive statistical package. However, there are some important typical differences. With the statistical package, the basic worksheet (the table of rows and columns of cells labelled A1, A2, etc.) usually contains only the data; the commands to produce statistics, and labels to describe the meaning of the variables plus the values that variables can take, are normally written outside the basic table (or *matrix*) of data. In a typical spreadsheet, the cells of the worksheet may contain any combination of data, statistical formulae, and labels to describe the data. More importantly, when a formula is typed into the cell of a spreadsheet, what actually appears in the cell is the answer—i.e. the result of the formula working on the data, not the formula itself. The actual formula will normally be displayed in the *command area* above or below the worksheet proper. So, the spreadsheet gives instant answers when formulae and other commands are typed in.

Even better, the spreadsheet is supremely tolerant of mistakes! If we discover that some of the data entries are wrong, we simply take the cursor to the wrong entries and, using the 'edit' facility of the spreadsheet, we overtype these cells with the correct values. Without further prompting, all the formulae entered into the worksheet recalculate and display results for the modified data! There is, therefore, no need to command recalculation of results for the good data; all this happens automatically. The formulae that we entered into the worksheet have been on permanent stand-by, ready to go instantly into action to produce updated results. Finally, as with statistical packages, the data and results of our manipulations can be stored for future retrieval and re-use.

Figure 3.1 above contains a mixture of what we typed, plus the routine row and column labelling (by the spreadsheet package) of the cells in the worksheet, plus the calculations performed by the package. What we actually keyed into the cells of the worksheet to get the resulting screen display reproduced as Fig. 3.1 is shown below in Fig. 3.2.

FIRM	Employees	Profit92	Profit93	Both92+93	Per Employee
SFB plc	12	23000	18900	@SUM(C3..D3)	+E3/B3
Teeswork	45	1200	23000	@SUM(C4..D4)	+E4/B4
Autocan	108	−25000	−45078	@SUM(C5..D5)	+E5/B5
NHFG	467	−135900	678000	@SUM(C6..D6)	+E6/B6
Cleveway	25	140000	−2300	@SUM(C7..D7)	+E7/B7
Easterside	67	−36790	1500	@SUM(C8..D8)	+E8/B8
TOTAL	@SUM(B3..B8)	@SUM(C3..C8)			
MEAN	@AVG(B3..B8)	@AVG(C3..C8)			

Fig. 3.2 Spreadsheet example worksheet (as keyed)

Note that columns E and F (the final two columns) were calculated by the spreadsheet from the formulae that we supplied. The formulae in column E ('Both92+93') use a 'built-in' function (@SUM) supplied by the package. Column F ('Per Employee') calculates the combined 92/93 profit per employee using some simple arithmetic supplied by the user (the '/' symbol means divide). The totals and arithmetic means (averages) for columns B and C are also obtained via formulae.

Generally, users like the immediate and intuitively familiar worksheet style of the spreadsheet. However, since data, formulae and text (labels) can all be mixed in the worksheet, more care has to be taken in manipulating the varied contents of cells. Being general tools, spreadsheets usually offer many fewer pre-written purely statistical routines than do dedicated statistical packages. This means that spreadsheet users have to work harder in turning statistical formulae from books into a command form that the spreadsheet will understand.

Do not worry if you do not have access to all the types of computer and package described. When a worked computing example is provided, try to reproduce it exactly (if you have the same hardware/software combination) or else try it (with suitably modified commands) on the package and computer that you use. If you can get the same answer then obviously all is well. You can then move on to some of the other statistical examples that we have not reworked by computer.

Using the computing examples

If you are not currently working with a computer, there is no problem. The obviously computing material in the book can be passed over until you find you need it.

Table 3.1 below indicates with the asterisk or star symbol (★) which packages feature in each chapter. All computer users should read Chapter 4.

Table 3.1 Packages featured in each chapter

	3	4	5	6	7	8	9	10	11	12	13	14	15	16	17	18	19	20	21
MINITAB	★	★	★	★	★	★		★	★		★	★		★	★	★		★	
SPSS^x		★	★					★	★	★		★		★	★				
SPSS/PC+		★	★							★	★		★		★	★			
Spreadsheets	★	★					★			★	★						★	★	

Readers will find regular cross-referencing between packages and chapters. For instance, the SPSS^x procedures introduced in Chapter 10 are repeated for SPSS/PC+ in Chapter 11. SPSS/PC+ users are advised to read Chapter 10, even though it focuses mainly on SPSS^x, in order to get to know the style of SPSS software. Limitations of space prevent the treatment of other well-known packages like **P-STAT** and **SCSS**.

4

Data entry and the computer

Importance of data entry

Despite the discussion in Chapter 2 of more sophisticated forms of electronic data collection, by far the most common approach for business surveys is to collect the data on paper and then transfer them to a computer file for statistical processing by the sort of package introduced in Chapter 3. The present chapter concentrates on the need to enter data as accurately and smoothly as possible into the computer. Some of the procedures described may seem tedious and bureaucratic. But that is the nature of large parts of research!

That stage of research which involves getting the data into the computer and then cleaned of 'bugs' (errors), is often not explained to students and newcomers to research. This is a pity since *data entry* and *validation* are key parts of an enquiry. They are stages that typically take a lot of time—much more time than researchers and student project writers normally anticipate! They are also stages that hardly ever fail to benefit from careful long-range planning and meticulous procedure and record keeping. There is little point in having acres of data if such data cannot be got cheaply, efficiently and accurately into a computer datafile for analysis. Inaccurately recorded data are bound to give inaccurate statistical results! Computer scientists have a word for it: 'GIGO'—garbage in, garbage out!

Packages and files

The computer has the advantage that it enables us to analyse easily very large *datasets*—perhaps data from hundreds of firms, or from thousands of citizens. The dataset needs to be keyed in only once. The data can then be stored electronically as a file within the computer, or on floppy disk or diskette (in the case of a small PC). We can return to the data file time and time again for further analysis, just as we might return to a file of papers stored in a conventional metal filing cabinet.

Our analysis can only be as good as the quality of the data introduced into the computer and stored there. To minimize the risk of clerical errors and misunderstandings in transferring the data from questionnaires, interview schedules, report forms and other sorts of paper documents, it is worth planning data collection and data entry into the computer carefully, and then checking the success of the data entry. Long before the researcher takes to the field, we need to consider how the data is to be entered into the computer. If data is to be gathered via an interview schedule or self-complete questionnaire, then the design of the survey instrument should take account of the needs of data entry and of statistical processing. Why this is so is dealt with in the middle part of this chapter. However, before we take a look at the best ways to plan the entry and storage of data we should have a clear idea of our goal. What will the stored data look like once it is in the computer?

In general, most survey replies are turned into numerical codes or scores, i.e. into an apparently 'quantitative' form. There are now some packages that are designed to analyse 'qualitative' data—i.e. free format verbatim replies by respondents to open questions—but these packages are not mainly designed for statistical analysis of the sort that this present volume deals with.

The data table

Statistical analysis by computer conventionally requires that replies are turned into figures. Fortunately, computer software appears to store and use data just as the lowly human with paper and pencil might set about dealing with household accounts—in rows and columns of figures. Ultimately, therefore, we want to store the data within the computer in a file which takes the form of a large *matrix* or table. Figure 4.1 would not be an unusual example, especially of a big dataset stored on a mainframe computer for analysis by a package like SPSS^x (see Chapter 3).

Columns . . .

```
00150213004217763418840012475639912218732098
00272100008119986210036292184423199475540000
00363317115119854678234419843977756399722109
004 . . . . . . . . . . . . . . . . . . . . . . . . . . . . . . . . . . . etc.
005 . . . . . . . . . . . . . . . . . . . . . . . . . . . . . . . . . . . etc.
  .
  .
  .
etc.
```

Rows

Fig. 4.1 Example of a simple dataset stored as a computer file

Notice in the above figure that there are no labels or explanations for the data in the computer file. Depending on the package we use, we can add names and labels later. In Fig. 4.1 each row represents one unit from our sample of respondents, or firms, or households or departments—whatever type of survey unit we have researched. In this example, each unit or 'case' takes just one row. Each row or case has an identifying number (001, 002, etc.). A three-digit number is chosen for the first three columns because there will be more than 100 cases or units for which data have been obtained. For each case or row, the columns beyond column 3 are occupied by the information gathered from a variety of questions or 'variables' (see Chapter 5 for a fuller treatment of variables and attributes). Note that in Fig. 4.1 there are no spaces between any of the columns of data. In this example, therefore, it will be left to us to tell the statistical package where the information for each variable starts and ends—i.e. which variables occupy which columns. Fortunately, some statistical packages make this process of data definition very easy! Readers may take comfort in the thought that the highly compressed form of the data table in Fig. 4.1 can easily be avoided by introducing blank spaces between each variable, for example:

001 50 2 1300 42 1 77 6 3 . . .

However, this requires more storage space in the computer.

Codes and levels of measurement

First we must decide which columns represent which variables. For some variables the meaning of the figures is obvious—a respondent's year of birth can be recorded as 1950 or as 1972 with no need for elaborate explanation. In practice, we may shorten the year of birth to 50 or to 72 with little fear of difficulty (columns 4 and 5 in Fig. 4.1). Recording monthly salary (columns 7–10) or weekly pay similarly presents no data 'coding' problem. Weekly pay is measured in an obvious way at what is described as the *ratio level of measurement*. At the ratio level of measurement the intervals between units of measurement are constant, and the zero point in the scale actually means zero or nothing. It is perfectly possible to earn zero or nil pounds per week (see case 002 in Fig. 4.1). Eventually, we shall discover that statistical manipulation of data at the ratio level of measurement can be very sophisticated.

Some variables that we might want to record in business (e.g. the maximum and minimum temperatures at which a batch of test portable tape recorders will function properly) are measured on familiar scales (e.g. degrees Celsius), but at the lower *interval level of measurement*. At the interval level of measurement, the intervals or steps between each unit of measurement are constant (i.e. the step from 12 to 13 degrees is the same magnitude as the step from 24 to 25 degrees), but the zero point on the scale is arbitrarily chosen—the zero point does not mean 'no temperature'. The arbitrariness of the zero point in temperature scales can be seen by comparing zero on the Celsius and Fahrenheit scales. Zero on the Celsius scale is the freezing point of water. Zero on the Fahrenheit scale is well below the freezing point of water (32° F)! The arbitrariness of the zero point on the interval scale of measurement means that such statements as '24° C is twice as warm as 12° C' are quite meaningless.

To prepare the data for statistical analysis we must always be aware of the level of sophistication of the information gained about cases. For many variables the coding selected in order to enter the data conveniently into the computer is a matter of convention. We might, for example, have conducted a survey of employed and unemployed business managers, asking them how strongly they agreed or disagreed with the statement: 'In the next six months the national economy is likely to expand'. We have asked a deliberately vague question, offering no definitions of what we mean, because we are anxious to tap the basic 'gut' reactions of business managers. Our sample of managers has been given seven possible response categories:

Could not agree more	Strongly agree	Mildly agree	Neither agree nor disagree	Mildly disagree	Strongly disagree	Could not disagree more
7	6	5	4	3	2	1

We have asked our respondents to circle the response category that best gives their opinion. The categories take the form of a simple scale running from 'could not agree more strongly' through the neutral 'neither agree nor disagree' to the other extreme of 'could not disagree more strongly'. Note that we have chosen to label the points of the scale from 7 to 1, and it is these 'scores' or codes that will be entered as the respondent's reply to this question. These scores are purely a convention. We could have chosen to label the seven scale positions in the reverse direction—i.e. from 1 to 7. Similarly, we might have chosen to label the positions from +3 through 0 (zero) to −3.

When we handle replies to this question at the analysis stage, we shall have to bear in mind how we decided to score the reply categories. We should remember that the scale is at the cruder *ordinal* level of measurement. This means that the scale is simply one of categories ranked or placed in order, and that we have no way of knowing whether the seven scale positions are equally spaced. The scale is like the results of a horse race where we know only which horse came first, which second, etc. We should certainly not assume *equal* spacing, or even that all respondents saw the scale in the same way. We should certainly not fall into the trap of claiming that the 'could not agree more strongly' reply is 'seven times more positive' than the 'could not disagree more strongly' reply. Our scores or codes remain a simple convention, designed to make data entry and storage simple and efficient.

The arbitrariness of coding for variables is best seen with variables (or more exactly 'attributes') at the *nominal* level of measurement. In our survey of managers we might code personnel managers as '1', production managers as '2', marketing managers as '3' and so on (column 6 in Fig. 4.1). In this case there is no claim whatsoever of measurement or ranking. The numbers are simply used as labels for groups, in much the same way as we number houses in sequence in order to identify them within the street, without any pretence that a 'number 6' house is worth three times as much as a 'number 2' house. Some variables are very simple, but likely to be important in analysis—e.g. sex. Sex is a *dichotomous* variable; it can take only two values. It is of no consequence whether we code men '1' and women '2' or vice versa. No idea of ranking or value is implied. The code is a pure convention. Other important variables, like occupation or social class, may be highly complicated—either because the categories can be counted in the hundreds (occupation), or because classification is problematic (social class). Fortunately, in these cases there may be ready-made coding schemes available, e.g. the occupational codes and class codes of the Office of Population Censuses and Surveys' *Standard Occupational Classification, Vols 1 & 2* (London: HMSO, 1990). An earlier version of this reference work was known as the *OPCS Classification of Occupations* (London: OPCS, 1980).

THE CODEBOOK

Data prepared for keying into the computer has to be well managed. We should keep a separate record of which variables will appear in which columns of the computer file, and what each of the numerical codes means. This record is known as the *codebook* or *data dictionary* and is a crucial document. The codebook will also have to record how we are to deal with 'don't know' or 'does not apply' replies to survey questions, or to data which are simply 'missing'.

Some statistical packages (e.g. MINITAB) specify in advance how missing data is to be coded. Other packages (e.g. SPSS[x]) allow the researcher to specify the codes. This gives the researcher more work, but also allows greater flexibility of procedure at the analysis stage. In our fictitious management questionnaire example, we might decide to code the 'economic prospects' question from 7 ('could not agree more') to 1 ('could not disagree more'). We would then use '8' to code for 'don't know' and '9' for 'missing data', and we might choose to reserve '0' (zero) for 'not applicable'. All these choices of codes will have to be recorded in the codebook. It is a good idea, wherever possible, consistently to reserve the same codes for the 'don't know', 'missing data' and 'not applicable' replies to all questions. So, all 'missing data' will be coded as '9' or '99' or '999', depending on the column width of the particular variable. Similarly, the code for 'don't know' for all variables might be '8' or '88', and so forth. This consistent approach will tend to reduce errors due to confusion in the coder. Using a consistent code is less taxing than having to remember separate codes for each variable.

In some circumstances we may not want, or be able, to distinguish between 'missing data' and 'don't know'. Where this is true both may receive the 'missing data' code in the codebook. The first few lines of a codebook for the example in Fig. 4.1 might look like Fig. 4.2.

Figure 4.2 illustrates how a codebook might be drawn up. We set out the variables against the columns that they will occupy in the computer dataset, giving details of all the codes used. We have also given each variable a short unique name. This will be useful if we are later able to introduce variable names into the computer file (most statistical packages allow this refinement), and is always useful in the interests of precision where a single question generates more than one variable (see opposite). Figure 4.2 also illustrates how, for some variables, further definition has to be recorded (see variable recording monthly salary (MONSA)). The interview schedule or questionnaire will have specified that the information be given in this way—monthly individual salary, after stoppages, as an average (mean) for the last three months.

CODING

As the researcher works through each questionnaire, the codes which correspond to the respondent's reply to each question can be recorded in one of several ways.

One possibility is to dispense with separate coding sheets, and instead incorporate the coding directly in a specially reserved portion of the

Column	Question number	Variable description	Codes	Name/Note
1–3	—	Respondent identifier	001 onwards	ID
4–5	1	Year of birth	99=Missing data (inc. refusal)	BIRTH
6	2	Management speciality	0=Generalist 1 Personnel 2 Production 3 Marketing 4 Finance 5 Distribution 6 Other 9 Missing data	SPEC
7–10	3	Monthly salary	Actual amount (£) 9999=Missing data (inc. refusal)	MONSA Net of tax, NI pension. Individual (not household). Average (mean) last three months.
11	4	National economic prospects	0=Not applicable 1 Could not agree more 2 Strongly agree 3 Mildly agree 4 Neither agree nor disagree 5 Mildly disagree 6 Strongly disagree 7 Could not disagree more 8 Don't know 9 Missing data (inc. refusal)	NATPRO

Fig. 4.2 Codebook for the first five variables (11 columns) of the data in Fig. 4.1

questionnaire or interview schedule. This is why it is suggested that the codebook should be written at an early stage, even before the questionnaire or interview schedule reaches its final version ready for printing. Obviously, the codebook must be settled well in advance if its most immediate practical outcome—*the coding frame*—is to be incorporated into the research instrument itself. The coding frame is the set of boxes into which the codes will be entered. Most of us are familiar with reserved sections of government or business forms headed 'For Official Use only'; the research procedure is exactly the same.

In surveys where the interviewer or the respondent is expected to write nothing but simply to tick boxes or circle pre-written replies, then it may be possible to dispense with the coding boxes and coding stage entirely. In this case, the numeric codes are already incorporated into the boxes designed for interviewer or respondent use. Later, the keyboard operator simply keys the code of the chosen boxes directly into the computer data file. Fig. 4.3 illustrates a few lines from this type of questionnaire.

Please tick (√) appropriate box in each question:

1. Are you male or female?   Male | 1 | Female | 2 |

2. In which department of the factory do you work?

Production | 1 | Maintenance | 2 | Dispatch | 3 |

Fig. 4.3 Example of a questionnaire with the response boxes incorporating codes

Data entry   *Data entry* is the process of getting the data into a computer, to be stored as a *file* (just as we store papers in a cardboard file). Once it has been decided how the coding will be undertaken, once the coding instruments have been drawn up, and once the coding has actually been accomplished, the next task is to key the data into the computer. Today this would probably be done in one of two ways. Where we are using a large mainframe computer (i.e. not a self-contained desktop machine), and where we are intending to use a non-interactive statistical package like SPSS^x (see Chapter 3), it is most likely that we would input data to the computer via the computer's *editor*. The editor is a piece of software, separate from the statistical package, which allows the input of programs, data or text into the machine. The editor can be thought of as a simple wordprocessor with which we can introduce data into the computer and later amend it. Differently from most sophisticated wordprocessors, the editor will probably require us to press the ⟨**RETURN**⟩, ⟨**Enter**⟩ or ⟨↵⟩ key at the end of each line or row of data. The file created must then be given a name. The name we choose for the file should give some indication of the file contents. The data file illustrated in Fig. 4.1 might well be input via a mainframe editor and be given a file name such as 'MANAGE1'. Comic file names such as 'FRED' should be avoided. Even a few days later it is difficult to remember, without a clue in the name, what the FRED file contains!

An alternative method for entering data is to use the statistical package's own built-in editor. Statistical packages with a more interactive style (i.e. user commands are executed directly on input from the keyboard, and results are immediately displayed on the screen) are likely to have a simple editor dedicated to the task of data entry and modification. The actual details of operation will vary between packages, and between mainframe and micro-computer versions of the same package. To some degree the choice between general editor or dedicated package editor may be one of personal preference and computer familiarity.

Whichever method we use, we should arrive at a set of data stored electro- **Storing data**
nically within the computer or on a storage device like a floppy disk or diskette.
The file will have been given a name.

As we input the data we shall have 'saved' the file every few minutes—i.e. the
computer is told explicitly to store everything that we have typed in up to that
point. If we wait until the very end of data entry before issuing the save
command there is the serious danger that some small error on our part, or a
computer crash beyond our control, will prevent us from storing all our hard
work. If this happens we shall have to re-key all the data. If we save every few
minutes the worst that can happen is that we lose a few lines of data, not the
entire file. This precaution is particularly important if we are new to the machine
or to the software.

At this stage, even if the statistical package offers the facility, it is probably
best not to attempt to add variable names and value labels (i.e. what the stored
codes mean for each variable) to the file. The addition of such stored names and
labels typically greatly complicates the way in which the dataset is stored by the
computer, thus making the process of checking the data more difficult. The
variable names and value labels can be added later, when we are satisfied that
the data have been fully checked and corrected. It is to the checking or
validation of data that we must now turn our attention.

Validation is the process of checking and correcting data files. It is highly **Data**
likely that the two processes of data coding and keying in will have introduced **validation**
errors. These coding and keying errors are in addition to sampling error and
other problems such as 'interviewer error'. If errors are small and random,
resulting inaccuracy is annoying and disappointing, but is unlikely to invalidate
our results. However, if errors are large and/or systematic, we have grounds to
worry; we must not analyse the data before the major coding and keying errors
have been discovered and corrected.

Once the data have been entered into a computer file, the file can be printed
out from the computer onto paper. This simple procedure allows us to take the
paper record and study it at our leisure in a search for errors. We can make the
obvious check that all data start and finish where they should—that there are no
empty columns where data should be, and no data spilling out into columns that
should be empty. Other easy validation techniques are available to us beyond
visual inspection. For example, we can use the statistical package to generate
from the entire dataset a one-way table for every nominal or ordinal level
variable, plus a report on the maximum and minimum values for every interval
and ratio level variable. If we discover any instances which are 'out of
range'—i.e. occurrences of theoretically impossible or unlikely codes—then we
have identified some sort of miscoding or miskeying error.

Some statisticians like to check data, and to start getting immersed in the
character of the data, by calculating simple statistics for just the first five or six
cases. They do this twice—once from the original research documents by simple
counting and the use of a hand calculator, and once using the statistical package
to process this small subset of cases in the computer dataset. If the two sets of

results, i.e. from hand processing and computer processing, do not agree then there is a problem which must be investigated further.

Another validation technique tries to find instances where miscoding or miskeying has resulted in cases that possess 'impossible' combinations of attributes or variables, even where the individual variable codes are *apparently* legal. The validation technique therefore lies in a cunning choice of variables to be cross-tabulated. In a household survey, for example, we might have asked men and women about their family histories. At the data validation stage we may ask the statistical package on the computer to generate from the data file a bivariate (two-way) table of the variable 'sex' by the variable 'number of children born' (i.e. a question that should only have been asked of women respondents). If the table generated by the package seems to have found men who have borne children, we have discovered a problem.

Back-up

Back-up is the copying of files to overcome accidents! When all the data validation has been completed, back-up is a further essential task. Data files can become *corrupted* in use or in storage and we should therefore immediately make at least one back-up copy of the data file. This will be of crucial importance in student projects! Diskettes, especially of the floppy variety, are easily lost, scratched or bent. Once the disk is corrupted the data file it contains will be inaccessible, and all the hard work of data entry and validation will have to be repeated. Even hard disks (storage devices within the computer) can become corrupted, or the files held in them may be accessible to mischief in an open environment such as a college computer laboratory. Alternatively, we may make some terrible error resulting in file deletion or corruption in the middle of a statistical analysis! It is therefore important to take out the simple insurance policy of storing the cleaned data file several times (i.e. under separate names), possibly in different, independent media (hard disk, floppy disk, magnetic tape).

Fortunately, most well-run mainframe computer systems automatically create back-up copies of files on regular (usually daily) 'disaster saves', but the micro-computer user will have to be more active in thinking of back-up needs. Whatever the system used, it would be foolish not to have second and third copies of the dataset stored safely away.

5

Tabulation

Method

Classification

In Chapter 2 we saw the various methods of collecting data, but before it can be tabulated, interpreted and presented in its final form, it must first be classified.

Classification is the process of relating the separate items within the mass of data we have collected. Every piece of data has its *characteristics*. Data collected about people may have, for example, characteristics concerning age, sex, height, weight and occupation, i.e. features which characterize the data and make it possible for us to classify it under particular headings.

Characteristics fall generally into two classes—*measurable attributes* (*called variables*) and *non-measurable attributes*. Non-measurable attributes are those attributes of the data which are not ordinarily measurable in units, e.g. disease, sex, colour, beauty. Variables are those attributes which are measurable, e.g. height (centimetres), weight (kilograms), absences from work (days), population (numbers), etc.

Discrete and continuous variables

Variables can be further subdivided into discrete and continuous variables. Discrete variables are those which can be measured only in single units, e.g. numbers in the population, houses in a town, tractors produced, size of clothing, etc. With this type there are gaps throughout the whole range of values that are not occupied by any items at all—e.g. shoe sizes, where the values jump from, say, $6\frac{1}{2}$ to 7 and there are no items of, say 6.73.

Continuous variables are those which are in units of measurement that can be broken down into infinite gradations, e.g. temperature (decimals of a degree), height (decimals of a centimetre), etc. In theory, this is always so, but in practice the gradations may be rounded off. For example, it is possible for two people to differ in age by one second, but their birth certificates will merely indicate that they were born on the same day.

This may seem rather surprising and confusing when we recall seeing in the press such statements as: 'The average number of children per class in England and Wales is 35.48.' One cannot have 0.48 of a child and, for classification purposes at least (!), the number of children is regarded as a discrete variable.

Tabulation   This is the process of condensing classified data in the form of a table so that it may be more easily understood, and so that any comparisons involved may be more readily made.

Simple   Suppose that a survey is carried out in a firm to investigate the absences, over
frequency   a period, of 200 workers, male and female, on the production line.
table   A simple survey of the firm's records could be used to collect the information needed, and this might include the following: name, sex, absences, cause of absence and nature of illness.

A summary table of the days of absence through illness for each worker might be compiled as follows:

Table 5.1 Number of absences (because of illness) for individual workers in a firm employing 200 in the production department

10	15	2	27	4	2	15	5	3	1	1	3	6	7
9	2	14	6	20	11	8	20	5	3	14	1	4	1
11	16	2	1	1	31	2	17	2	25	3	7	4	15
12	2	13	4	18	1	27	3	10	1	5	1	1	42
5	3	4	20	6	26	3	10	6	3	15	2	11	2
7	5	2	4	22	7	1	5	21	6	1	24	3	10

In addition to the above records, 116 workers had no absences due to illness.

In order to tabulate this data, we must first classify it by the variable 'number of absences'. This is an example of a *discrete variable*, because, by the method of recording absences used by the firm, any absence involving part of a day counts as a full day, no fractions of a day being recorded.

The classification of this data could be done by arranging the above figures into order, highest to lowest, e.g.:

$$42, 31, 27, 27, 26, 25, 24, 22, 21, 20, 20, 20, \text{etc.},$$

This arrangement is known as an *array*.

A less tedious method would be to count how many workers were absent for one day, two days, and so on. This might be done as follows by means of *tally marks*:

Table 5.2

Class interval	Class frequency	
Days absent	Number of workers involved	
1	ͰͰͰͳ ͰͰͰͳ 11	= 12
2	ͰͰͰͳ ͰͰͰͳ	= 10
3	ͰͰͰͳ 1111	= 9
4	ͰͰͰͳ 1	= 6
etc.	etc.	etc.

When this tallying has been carried out, we should have a long list showing how often workers are absent, with a range from one absence to 42 absences. Such a table, however, would be too lengthy. It would contain many zero frequency entries (e.g. from 32 to 41 inclusive would all be zero!). The list would be wasteful of paper, unbalanced and not very easy to read.

Instead of using an *ungrouped frequency distribution* such as this it is usually more convenient (even if a little less precise) to make a *grouped frequency distribution*. In order to do this it is necessary to arrange the data into groups, or classes, into which the frequency of absence will fall. Such classes are known as *class intervals*. The first step is to find the *range*. This is simply the difference between the highest and the lowest values of the variable (number of days absent).

In our example this would read:

Highest variable value = 42 days
Lowest variable value = 1 day
Range = 41 days

We must use our discretion in choosing the size and number of classes in which to split the range. It is wrong to make the class intervals too small, because this would result in a lengthy list of intervals and would have the same fault as the ungrouped frequency distribution. It is also wrong to make the class intervals too large, because this would result in only a few class intervals, and a good deal of precision would be lost. The assumption is that the frequencies within an interval are distributed equally throughout it.

Wrong	*Right*	*Wrong*
1 to 2	1 to 4	1 to 9
2 " 3	5 " 9	10 " 19
3 " 4	10 " 14	20 " 29
etc.	etc.	etc.

The number of class intervals in a usual frequency distribution is hardly less than 5 and usually not more than 15. In the wrong examples above we would have had 41 intervals in the first case, and 5 in the second case. In the right example we would have a table as follows:

Table 5.3 Absences due to illness in production department

Absences (days)	*Number of workers*
1 to 4	37
5 " 9	17
10 " 14	11
15 " 19	7
20 " 24	6
25 and over	6
TOTAL	84

Each class interval gives the number of days (inclusive) of absence, i.e.:

1 to 4 includes 1, 2, 3 and 4
5 to 9 includes 5, 6, 7, 8 and 9, etc.

The last class interval is an example of what is called an *open-ended class interval*. This is a device often used when the items for inclusion are few and widespread (in our example, only 6 items covering a range of 25 to 42). To have continued the list of class intervals to encompass the few items involved would have meant lengthy and unproductive work. Such open-ended class intervals may occur at the beginning or at the end of a frequency distribution, for example:

149 and under
150 to 159
160 " 169
etc.

Complex frequency table

In Table 5.4 below, our previous table has been further subdivided, and we have compared *attributes* (male and female), as well as simply tabulating the frequencies of the *variables*.

Table 5.4 **Absences due to illness in production department**

	Number of workers		
Absences (days)	*Males*	*Females*	*Total*
1 to 4	19	18	37
5 " 9	9	8	17
10 " 14	5	6	11
15 " 19	3	4	7
20 " 24	2	4	6
25 and over	5	1	6
TOTALS	43	41	84

In the next example (Table 5.5), the number of absences due to illness has been further subdivided into 'Accidents at work' and 'Other causes'.

This is a fuller presentation of the complex frequency distribution. The percentage absences in the last row are written in italics to make them stand out from the absolute figures in the table, and they are placed near the totals to which they relate. The footnote on the meaning of 'days' is included for the reader's information, as are the sources of the information.

Table 5.5 Absences due to illness in production department, January to December, Year 1 (AGP/Wages Dept. and Health/Personnel)

Absences (days)*	Number of workers								
	Males			Females			Totals		
	Acc. at work	Other	Total	Acc. at work	Other	Total	Acc. at work	Other	Total
1 to 4	4	15	19	6	12	18	10	27	37
5 " 9	3	6	9	3	5	8	6	11	17
10 " 14	1	4	5	2	4	6	3	8	11
15 " 19	0	3	3	1	3	4	1	6	7
20 " 24	1	1	2	2	2	4	3	3	6
25 and over	3	2	5	0	1	1	3	3	6
TOTALS	12	31	43	14	27	41	26	58	84
Total workers in department		80			120			200	
% absences		53.8			34.2			42.0	

* A fraction of a day is counted as a day.

Acc. = accidents.

A slightly different type of table can be constructed as shown in Table 5.6 to give the frequency of absences at the various points of the class intervals:

Cumulative frequency table

Table 5.6

	Column 1	Column 2	Column 3
	Absences (days)	'Cum' less	'Cum' more
Row 1	1 to 4	37	84
Row 2	5 to 9	54	47
Row 3	10 to 14	65	30
Row 4	15 to 19	72	19
Row 5	20 to 24	78	12
Row 6	25 and over	84	6

'*Cum*' = cumulatively.

It is possible from column 2 in this table to find how many workers had, for example, less than 10 days' absence through illness. If we look along the second row of class intervals in column 2, the figure is 54. This cumulative table is calculated simply by adding together the successive totals for each class interval in order, from the lowest to the highest class interval. Similarly, in the 'cum' more column, column 3, we can find, for example, the number of workers who had 15 or more days' absence by looking at this column along the fourth row, where we find the answer, 19. This cumulative column is constructed by adding the successive totals for each class interval from the highest to the lowest, in order. In this case we enter the successive totals from the bottom upwards.

Other forms of tabulation are commonly met and are noted below.

Tabulation of time series

The tabulation of a time series is the record, over time, of how a variable has changed in value. For further information on time series, the reader is referred to Chapter 12. The table below gives an example of the tabulation of the growth of bank advances over the period Year 1 to Year 7:

Table 5.7 The growth of bank advances

Year (end December)	Advances (£m)	
	London banks	Scottish banks
1	4732	514
2	4725	503
3	5075	520
4	5328	548
5	5624	586
6	5991	634
7	9735	865

A distinguishing feature of a time series table is that columns are rarely totalled, but row totals are often used when data for each year are broken down into sections.

Tabulation by geographical location

Table 5.8 Fatal and non-fatal accidents, third quarter, by Divisions of Inspectorate (Department of Employment Gazette)

Division	Fatal accidents	Total accidents
West Riding and N. Lincs.	8	7771
Northern (Leeds)	12	6694
Midlands (Birmingham)	7	4565
Midlands (Nottingham)	13	5114
London (West)	9	4274
London (North)	6	3925
London (East)	7	4446
South Western	9	3150
Wales	11	4506
North Western (Liverpool)	10	5073
North Western (Manchester)	5	3871
Scotland	11	5922
TOTALS	108	59311

Location, either in time or space (geographical), is not considered as an attribute of an item. That is why this table, and the time series in Table 5.7 are not placed in the following section 'Tabulation by attributes.'

In these cases the characteristic is not measurable and it may refer to occupation, industry, personal details, etc.

Tabulation by attributes

Table 5.9 shows the causes of stoppage, i.e. strikes due to industrial disputes:

Table 5.9 Stoppages of work (Department of Employment Gazette)

Principal cause	Number of stoppages	Number of workers directly involved (hundreds)
Wages—claims for increases	1176	8395
Wages—other disputes	240	588
Hours of work	40	81
Employment of particular classes of persons	423	2592
Working rules and discipline	355	912
Trade union status	75	227
Sympathetic action	36	281
TOTALS	2345	13 076

The table below shows an analysis by industrial sector:

Table 5.10 Loans and overdrafts in Great Britain*

Industrial sector	Per cent at August 19—
Agriculture and fishing	4.4
Mining and quarrying	1.4
Manufacturing industries	32.1
Building and contracting	5.7
Finance (incl. H.P. companies)	20.6
Personal and professional	16.1
Services	19.7
TOTAL	100.0

*Advances by members of British Bankers' Association through offices in Great Britain: classified according to the business of the borrower – not the use to which the credit is put.

A feature of this table is that, for quick reference and to help comparison, all of the data have been reduced to percentage form.

By its nature, tabulation is extremely flexible and, after a point, it becomes difficult to lay down hard and fast rules. It is always up to the student to exercise his or her discretion in order to present tables in a clear, intelligible and attractive form. Nevertheless, there are certain basic rules and hints which should be noted in any tabulation, even of the more complex types:

Rules of tabulation

41

1. All rows and columns should be headed with clear, explanatory titles and a note of the units used.
2. Tables should bear suitable titles, and these should combine brevity with as full a description as possible.
3. Margins should be left round the edges of the table and, for neatness, the whole table should preferably be contained in a frame.
4. Double, bold and feint lines should be used to divide and sectionalize the data, especially where the table is complex.
5. Footnotes explaining points of classification, and any special notes, should be included with the table, as should a note of the source of the data.
6. Provision should be made for totals at the end of columns and rows, if this is necessary.
7. Comparative percentages, ratios, averages, etc., may be included, and should be placed close to the absolute figures to which they refer. (Such derived figures are usually printed in italics.)
8. Tables should not be overloaded and, if this is a danger, it is often better to separate the data into two or more tables.

Use and interpretation

Classification and tabulation

It is a good method to draft out the classification scheme and the tables to be used, before the actual survey is undertaken or the recorded data are collected. This is another example of planning the end before beginning with the details. If one proceeds in this way, many problems in the drafting of the questionnaire, e.g. deciding what units to use, may be avoided.

In the processing of the results, a check for 'errors of transference' between stages should be included, e.g. from questionnaire to classification, from classification to tabulation, etc. Unavoidable errors which may already be present in results, e.g. bias, should not be added to by errors within the statistician's control.

The actual analysis and tabulation of the results is, nowadays, often done by a computer if there are many items involved.

Drafting blank tables

From the methods described previously in this chapter, the student should be able to attempt the tabulation of a simple set of attributes. An illustration is given below of the possible steps in the drafting of a blank table.

Example:

A firm wishes to classify its employees according to type of employment, distinguishing between males and females, and showing trade union membership, if any. Rule up a blank table, with suitable headings and showing all necessary subtotals.

Note: The vertical divisions of a table are called columns, and the horizontal ones are called rows.

1. Make a list of the attributes and variables that you are asked to tabulate, with a note of the groups within each attribute, for example:

$$\left.\begin{array}{c}\text{Type of employment—4 groups} \\ \text{Sex—2 groups} \\ \text{Trade union membership—2 groups}\end{array}\right\} + \begin{array}{l}\text{totals and} \\ \text{subtotals}\end{array}$$

2. Divide the attributes between the columns and the rows, so as to give a well proportioned table, for example:

Sex and Trade union membership in columns
Type of employment in rows

Although a table is rectangular, we can only use two sides. The base and the right-hand side of the rectangle are normally used for totals, across and down.

3. Now rule up a rough table. This will probably need amending later, so allow plenty of space, and don't worry about details at this stage. The table might look as follows:

Table 5.11

Employment	Males		Females	
	Union	Non-union	Union	Non-union
Skilled				
Unskilled				
Sales				
Administration				

4. Totals and subtotals may now be inserted at the ends of columns and rows.
5. Now draft the final table, adding any refinements.

Table 5.12 Distribution of employees of Jones Co. Ltd by sex, trade union membership and type of employment, December 19—

Type of employment	Males			Females			Totals		
	Union	Non-union	Total	Union	Non-union	Total	Union	Non-union	Total
Skilled factory									
Unskilled factory									
Sales force									
Administration									
TOTALS									

Points to watch:
(a) Note the title, statement of units and date.
(b) Note the provision made for check totals at the bottom right-hand corner.

    (c) Note particularly that any subtotal can be easily extracted. This is the final test of a draft tabulation.

    (d) The commonest fault is the omission of some subtotal. The rule is that, once a sequence of headings is established (e.g. union, non-union, total) this sequence must be repeated in the final totals columns. A similar rule applies to rows, if there is any subdivision of these.

Class intervals

    In the Method section we dealt with class intervals when the variable was a discrete series. When the variable is a *continuous* series certain problems arise. These concern the meaning and the limits of the class interval. For example, if we were presented with class intervals of the age of the population, these might appear as follows:

> *Ages (years)*
> 0–4
> 5–9
> 10–14, etc.

The limits of the intervals given are: *lowest*—the first value in each interval; *highest*—the highest possible value before the lower limit of the next class interval. Strictly speaking, 0–4 means '0 years to 4.99 (recurring)' years.

    This explanation is necessary in the process of tabulation because we may be undecided, when presented with a class interval of 0–4, in which class frequency to place an item value 4 years, 11 months, 27 days. As this is just under 5, the item goes into the first class frequency.

    It is also necessary to warn the student that any of the following descriptions of class interval may be encountered:

> 0 but under 5   0 to 4   0–   0–5
> 5 but under 10   5 to 9   5–   5–10

All these are variations on the above example, and are often the source of much confusion. The last example is particularly misleading and should never be used. For example, in which class frequency—first or second—should an item value 5 be placed?

    Classification should be carried out according to the limits given (except of course in the single case mentioned above) and, if the intervals are written:

> –5
> –10, etc.

then, unless there is evidence to suggest that this is merely another variation of 0–4, classification should be as:

> 0.01–5
> 5.01–10
> 10.01–15, etc.

Similar problems do not, of course, arise in the case of a discrete series, where items are placed in the class frequency the interval of which is bounded by the simple inclusive whole units.

Starting analysis with the computer

We start our computer analyses with a statistical package, MINITAB. Spreadsheet users will find that they can easily reproduce the example below in the spreadsheets **Lotus 1-2-3**, **AS-EASY-AS**, and **Works** by first reading and practising the final sections to Chapter 9 (Averages). Then return to this example.

Imagine that we wish to enter the most important data on absences from Table 5.5 above into a statistical package. We have chosen the MINITAB package for this purpose. MINITAB comes in versions for both mainframe computers and for micro PC-style computers. With a mainframe computer, and with a micro-computer in a shared laboratory setting, it is likely that the user will have to 'log in' to the machine—i.e. identity himself or herself to the computer with an 'ID' and with a secret password. It may also be necessary to move through the computer's facilities to reach the area where MINITAB work can be undertaken. This route will vary with each computer installation. There should be local instructions to guide you to this point.

On a PC-style computer the instruction for getting to the right *directory* (work and storage area) is likely to be:

```
cd minitab ⟨RETURN⟩
```

This instruction *changes directory* (cd) to a directory called 'minitab' where the package is stored.

When we are at the right point, whether we are using a mainframe or micro, we call up the statistical package by typing MINITAB and then pressing the ⟨**RETURN**⟩ key (also marked ⟨**Enter**⟩ or ⟨↵⟩):

```
minitab ⟨RETURN⟩
```

We see some starting messages, and we know that the package has been successfully activated when we see on the screen 'MTB >'. This is the MINITAB *prompt*, and it lets us know that the package is waiting for us to input commands where the cursor lies.

Our first task is to input the data. Inputting the data to the MINITAB worksheet (the table or matrix of empty cells in rows and columns) is more easily done if we are using a PC micro-computer. In the PC version of MINITAB we simply press the 'escape' key, ⟨**Esc**⟩. We immediately see on the screen the empty worksheet with rows and columns numbered. The ⟨**Esc**⟩ key is a 'toggle' switch—pressing it takes us back and forth between the worksheet, for data input, inspection and editing, and the command area of MINITAB (where we saw the MTB > prompt) where the statistical work is done.

Inputting the data

45

Once we see the blank worksheet we can simply key the data into each cell. The *active* cell—i.e. where the data will appear on the screen—is indicated by the coloured or shaded highlight. Once the data for a cell have been typed from the keyboard, we can either press ⟨**RETURN**⟩ or ⟨**Enter**⟩, or else move to the next cell with one of the cursor 'arrow' keys, e.g.⟨ ↓ ⟩. Mistakes on input are easily obliterated with the ⟨**Del**⟩ key, or (after the ⟨**RETURN**⟩ key has been pressed) by returning to the cell and overtyping the wrong data with the correct data. The new user should behave intuitively: the screen replaces the sheet of paper, and the coloured highlight indicates where the 'point' of our imaginary electronic pen is lying on the screen. Figure 5.1 below illustrates data being input to the MINITAB worksheet. The data entered is the essential data from Table 5.5—absences due to illness in a production department. Column C1 contains the classes for length of absence (days). When you reproduce this worksheet as an exercise for yourself, take care to begin the entry of data into cell C1 with a word like 'from'. If you begin C1 with a number, the package will be expecting a simple numerical variable and problems will arise. Columns C2 to C4 contain the data for males ('Accidents at work', 'Other') and females ('Accidents at work', 'Other'). All the 'Total' columns and cells have been omitted. We shall derive these, by MINITAB calculation, later on.

ABSENDAY	C1	C2	C3	C4	C5	C6	C7	C8	C9
1:	from 1 to 4	4	15	6	12				
2:	from 5 to 9	3	6	3	5				
3:	from 10 to 14	1	4	2	4				
4:	from 15 to 19	0	3	1	3				
5:	from 20 to 24	1	1	2	2				
6:	25 and over	3	2	0	1				
7:									
8:									
9:									
10:									

Fig. 5.1 MINITAB worksheet with key data from Table 5.5

The micro-computer version of MINITAB allows us more easily than the mainframe version to introduce columns of data (like C1) which contain alphabetic, as opposed to purely numeric, characters. In the micro version we can type them into the chosen cells, taking some precautions (see note on C1, above).

Both versions of MINITAB, however, allow the easy input of column names to describe the data. Figure 5.1 shows the name 'ABSENDAY' above the column C1. This is short for 'Absences (days)'; MINITAB limits us to names of eight characters. In the micro version of the package, while in 'data entry' mode (i.e. working with the worksheet on the screen), we have moved the cursor (highlighted bar) with the cursor keys to any cell in column C1 and then pressed the function key ⟨**F10**⟩ on the keyboard. This action gets us into a short dialogue with the package, using a *menu* (list of available options). As part of the dialogue we type our chosen name for C1 and press ⟨**RETURN**⟩ or ⟨**Enter**⟩. In the micro version, in data entry mode, we can also press the ⟨**F1**⟩ function key to get help. The ⟨**F1**⟩ 'help' facility is also presented as a menu. We use the cursor arrow keys, e.g. ⟨↓⟩, to move up and down the menu. We use the ⟨**RETURN**⟩ key to select from the menu the item on which we want help. We use the ⟨**Esc**⟩ key to move from the help facility back to data entry.

In the micro-computer version of MINITAB, when we have finished our data entry into the worksheet, we press the ⟨**Esc**⟩ key and this takes us back to the command area. We know we have arrived back when the worksheet disappears from the screen and the MTB > prompt and cursor reappear. (Mainframe users of MINITAB will not find data entry difficult. Mainframe users should jump ahead to the computing section at the end of Chapter 6 for details of the alternative MINITAB commands to input data.)

Assuming that we have input the data to the MINITAB worksheet, we now **Completing** proceed from the MTB > prompt in the same way, no matter which version of **the table** MINITAB we are using. Our task, using simple MINITAB commands, is to complete the table of results as displayed in Table 5.5. First, we assure ourselves that although we can no longer see the worksheet on the screen the data still exist, and in the right columns. To do this we type the command INFO and press the ⟨**RETURN**⟩ key. The package responds immediately by producing a summary of the data. This summary appears on the screen as shown below:

```
MTB > INFO
            COLUMN   NAME        COUNT
A           C1       ABSENDAY      6
            C2                     6
            C3                     6
            C4                     6
            C5                     6

CONSTANTS USED:  NONE
```

The summary indicates that there are five columns of data, with six data items in each column. Column C1 has a name, ABSENDAY, and is a column of text rather than a strictly numerical variable. (Here C1 contains the row designations like 'from 15 to 19'). That C1 contains 'alpha data' is indicated by the label 'A'.

We wish now to calculate the total number of males with absences due to accidents and other causes. We do this by keying two lines of command. We press ⟨**RETURN**⟩ at the end of each line:

```
MTB > LET C6=C2+C3
MTB > NAME C6 'MALETOT'
```

The first line is a LET command. It creates a new column or variable (C6) by adding the entry in each row of C2 ('Accidents at work') to the corresponding row entry for C3 ('Other'). The result will be the column of data headed 'Males: Total' in Table 5.5. The second command gives the new variable or column, C6, a name. The name is 'MALETOT'. Notice that we should surround the name with *single* quotation marks when we key it in.

We repeat this procedure for the female employees, combining their two columns of data (C4, C5) into a new column, C7, which adds the figures for the two types of absence ('Accidents at work', 'Other') for the six periods of absence. We name the new column 'FEMTOT':

```
MTB > LET C7=C4+C5
MTB > NAME C7 'FEMTOT'
```

If you have access to the MINITAB package you should repeat the procedure this far. You should then create, for yourself, two new variables or columns of data, C8 and C9, which give the total absences (men and women combined) for accidents at work (C8), and the total absences (men and women combined) for absences due to other causes (C9). Give both these columns names.

Our next task is to use the newly created C6 and C7 variables to arrive at a new column or variable showing the total of all absences, for both sexes, in each of the six periods of absence categories. Once more we use the LET command for the simple computation:

```
MTB > LET C10=C6+C7
MTB > NAME C10 'ALLTOT'
```

By this time, having created two new variables (C6 and C7) and then a newer variable (C10) from these creations, we may be losing our confidence that the data are all as they should be! We may wish to review the data. This is easily done. If we are using the micro version of MINITAB we simply toggle to the worksheet by pressing ⟨**Esc**⟩. To get back to the MTB > prompt we press ⟨**Esc**⟩ again. Alternatively, whichever version of MINITAB we are using, we can display the data on the screen by typing in a PRINT command. The form of the PRINT command, and the instant response of the package, are given below:

```
MTB > PRINT C2-C7, C10
```

ROW	C2	C3	C4	C5	MALETOT	FEMTOT	ALLTOT
1	4	15	6	12	19	18	37
2	3	6	3	5	9	8	17
3	1	4	2	4	5	6	11
4	0	3	1	3	3	4	7
5	1	1	2	2	2	4	6
6	3	2	0	1	5	1	6

We complete the original Table 5.5 by calculating the percentage of workers sometime absent from the department. We know from Table 5.5 that the total number of male workers is 80. We enter this single figure as a *constant*. In MINITAB, constants are stored as a 'K'. In MINITAB *syntax* 'C' is for *Column*, 'K' is for *Konstant*, 'E' is for *Either* column or constant!

```
MTB > LET K1=80
```

So, the constant 'K1' contains the value 80. Next we add, in column C6, all the male workers who have been at all absent from work. We do this with the familiar LET command, but using the built-in function SUM. SUM adds all the data entries in C6. The total is stored in a second constant, 'K2'. We display it with PRINT:

```
MTB > LET K2=SUM(C6)
MTB > PRINT K2
      43.0000
```

We are now in a position to calculate the percentage of male workers who have been absent:

$$100 \times \frac{\text{absent male workers}}{\text{total male workers}}$$

In our worksheet and MINITAB terms this becomes:

```
MTB > LET K3=100*K2/K1
MTB > PRINT K3
      53.7500
```

Our MINITAB result is the same as that given in Table 5.5, but to a greater number of places after the decimal point.

Note that to display the results of a LET command, we have to use a second command line with PRINT. Also note that in MINITAB terms, multiplication is indicated by '*', and division by '/'. Exponentiation is indicated by '**'. The command LET C6=C5**2 means that each entry in the C5 column will be squared and placed in the corresponding row of column C6.

Using MINITAB, complete Table 5.5 to give the female percentage absent, and the percentage absent for both sexes combined. (*Note*. You can create a total of six more constants 'K4'–'K9'.) Check your MINITAB results against the original table.

Finally, we exit from MINITAB with the command STOP, followed by the ⟨**RETURN**⟩ key:

```
MTB > STOP
```

The sad news is that when we exit like this our data will not have been stored in the computer. How to save and store data permanently is discussed in the next chapter. When we have exited from MINITAB with STOP, we return to the operating system prompt of the computer that we are using. We may well have to log out or log off from the computer before physically turning off the machine. Details of how to do this will vary between installations.

Fundamental tables

Understanding and creating tables like this is an essential part of statistics. Most important is the ability to draw up a *contingency table*. This is a table which records how many observations fall into a 'two-way' (or greater) system of classifications—e.g. 'the population broken down by age and sex'! The computing sections of Chapter 17 deal at length with commands in MINITAB, SPSS[x] and SPSS/PC+ to create such tables.

Being able to manipulate data via commands like LET is also a valuable facility. The final section of Chapter 15 shows other ways of manipulating or 'transforming' data with MINITAB.

Charts, diagrams and symbols

Method

These are usually drawn to represent the total number of items in a group, at any **Bar charts** point in time (as in Fig. 6.1). Note the guidelines, which help the eye to judge the relative heights of the bars, and the vertical scale.

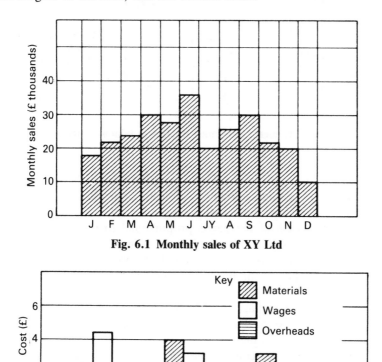

Fig. 6.1 Monthly sales of XY Ltd

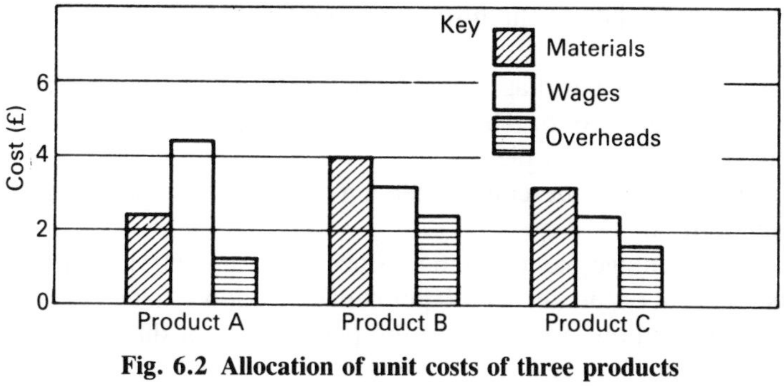

Fig. 6.2 Allocation of unit costs of three products

Sometimes two or more bars are drawn for each item (see Fig. 6.2). This is a *compound* (or *multiple*) bar chart. If the comparison is in percentages, the diagram is known as a *percentage bar chart*. Note that the bars are drawn vertically. There is no definite rule about this, but in the absence of any specific instruction by an examiner, the student should draw the bars vertically.

Finally, there are *component bar charts*. These are used to show the breakdown of a total into its component parts, and here again, we may show the actual figures (see Fig. 6.3), in which case the bars will be of varying heights, or we may show a percentage breakdown, in which case the bars will be of equal height, representing 100 per cent in each case.

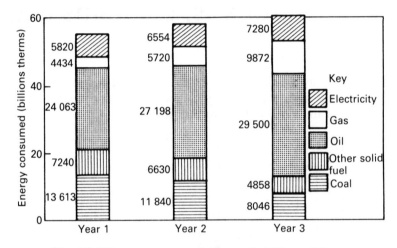

Fig. 6.3 UK energy consumed (figures in billions of therms)

Descriptions of the items involved may be written on, or beside, the bars, as may the actual or percentage amounts. Colouring or shading may also be used, but it is preferable to show the actual scale used in all cases.

In all cases, the *length* of the bar is proportional to the size of the item it represents, while the *width* of all bars in a diagram remains the same. Where axes of measurement are used, these should always begin at zero. A special application of a bar chart is the *histogram* (see Chapter 7).

Circular or 'pie' diagram

Like the component bar diagram, the 'pie' diagram can be used to represent the parts of a whole group. The different values of each item are drawn in proportion as the slices of a pie.

The pie diagram may be constructed on a percentage basis, or the actual figures may be used.

We have a percentage pie in Fig. 6.4. Since there are 360° in a circle, and this represents the total of the items (100 per cent), it follows that 1 per cent = 3.6°. The circle is marked out according to the percentages of the parts.

Pie diagrams may also be used to compare values in different years.

	Percentage
Wages	$33\frac{1}{3}$
Expenses	$12\frac{1}{2}$
Materials	25
Tax	$16\frac{2}{3}$
Profit	$12\frac{1}{2}$
TOTAL	100

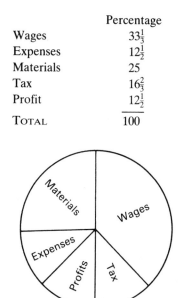

Fig. 6.4 Percentage outgoings in each £1 sales income

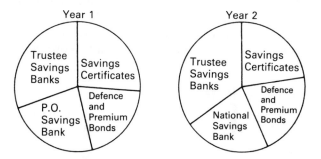

Fig. 6.5 Pie diagrams of National Savings

In Fig. 6.5 we have a comparison of National Savings in Years 1 and 2. In addition to working out the angles for the slices in each pie, we must make the pie proportionately larger. This is done as follows:

1. The respective totals (in £million) are (Year 1) £7786 and (Year 2) £9171. Therefore, the areas of the circles must be in this proportion, i.e. as 1 is to 1.18.
2. Since the area depends on the square of the radius (the area of a circle is πr^2) we must first find the square roots of 1 and 1.18. These are 1 and 1.09 respectively. We therefore draw circles with radii of (say) 2 cm and 2.18 cm.

3. We now split each pie according to the subdivisions. We give the figures below as an illustration:

	£m	
Savings Certificates	2032 =	94.0°
Defence and Premium Bonds	1610 =	74.4°
P.O. Savings	1779 =	82.3°
Trustee Savings Banks	2365 =	109.3°
TOTAL	7786 =	360°

The whole circle (360°) is represented by 7786. By simple proportion, 2032 is

$$\frac{360 \times 2032}{7786} = 94.0°$$

Similarly, 1610 is $\frac{360 \times 1610}{7786} = 74.4°$, and so on.

In general, comparison of circles of different sizes is best avoided, because the real variation is difficult to judge. For example, a circle which is twice the diameter is four times as large, but does not look so.

Pictograms   Sometimes these are known as *picturegrams*. An almost limitless variety of picture symbols may be used to represent values. As in the case of bar diagrams or pie diagrams, the picture symbols may show values at one point in time, or variation over time. A simple pictorial illustration is shown in Fig. 6.6.

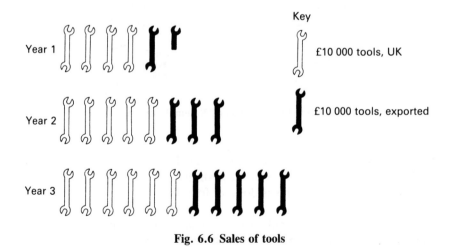

Fig. 6.6 Sales of tools

Each symbol represents a certain amount, e.g. £10 000 of tools is represented by a single spanner. Home sales and exports are distinguished by shading.

A defect of pictograms is their lack of precision, e.g. the sales can only be shown in units of £10 000. Occasionally, one sees a portion of a symbol drawn, as in Year 1, but there are obvious limits to this device. If greater precision is sought by letting each symbol represent a smaller amount, the number of symbols needed becomes excessive, and the eye is unable to take them all in.

A striking way of presenting geographical data by diagrams is the cartogram, as shown in Fig. 6.7.

Cartograms

This form of diagram demands, perhaps, more preparation, because a map has to be drawn and divided into the desired regions. Shading or hatching may be used, although frequently symbols can be placed according to the location of the item illustrated.

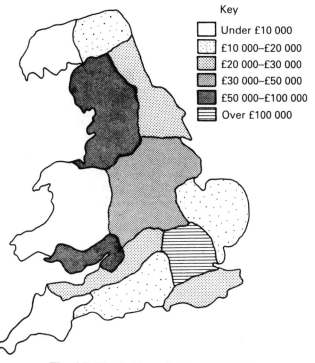

Key

☐ Under £10 000
▫ £10 000–£20 000
▨ £20 000–£30 000
▩ £30 000–£50 000
■ £50 000–£100 000
≣ Over £100 000

Fig. 6.7 Distribution of sales, XYZ Ltd

In presenting this kind of geographical diagram, it is helpful to provide a key, as in the cartogram above.

This type of diagram shows the simple percentage change in a series between two fixed points in time. It can be seen from the example in Fig. 6.8 that the 'base date' (earlier date) is regarded as zero on the vertical (percentage) axis. Straight lines are drawn radiating from zero to points which fix the percentage changes (+ and −) and the 'end date' (time on the horizontal axis).

Period change charts

55

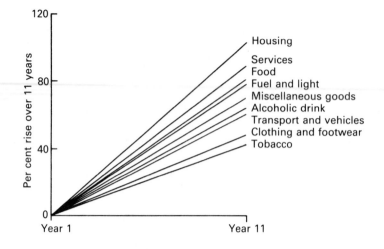

Fig. 6.8 Rise in retail prices, Years 1 to 11

Neither grids nor guidelines are used, because they might falsely show points of increase or decrease *within* the two dates. Axes should always be marked, but only the vertical axis need be calibrated, because the dates of the period are given by 1. the origin, and 2. the end of the change lines.

Scatter diagrams

Frequently we may wish to compare two sets of figures and to show how they vary with each other. For example, 10 small grocers' shops recorded the weekly figures of profits and turnovers as shown in Table 6.1:

Table 6.1

Shop	1	2	3	4	5	6	7	8	9	10
Profit (£s)	525	529	532	534	538	542	554	559	563	574
Turnover (£s)	2125	2170	2174	2180	2190	2215	2260	2300	2320	2400

We mark scales of profit on one axis, and turnover on the other.
Note. It is not necessary with this type of diagram to start the scales at zero, because we are only concerned with the *relative* positions of the points.

The points on our diagram are plotted by taking the profit and the turnover of shop number 1, and marking it by reference to the two axes. We proceed to plot the remaining points until a series of points is built up as in Fig. 6.9. The resulting cluster of points will show the relationship (or, more properly, the correlation) of profits to turnover in these shops.

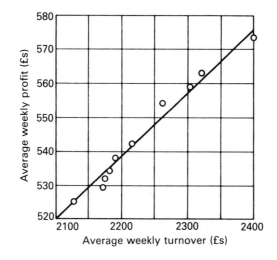

Fig. 6.9 Profit and turnover of 10 grocers' shops

If the points tend to lie along a line (as in Fig. 6.9), there is said to be a *strong linear correlation*. Also, as it appears in the diagram that profits increase as turnover increases, this is called a *strong positive correlation*.

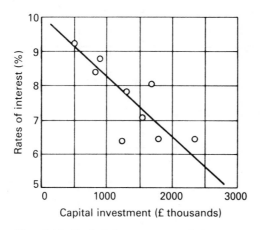

Fig. 6.10 Capital investment and interest

In many cases, however, we may have *negative correlation* where, as one item increases, the other decreases (see Fig. 6.10). Finally, we may have *weak correlation* (Fig. 6.11), or *no apparent correlation* at all (Fig. 6.12).

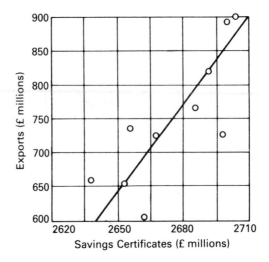

Fig. 6.11 Savings Certificates related to exports

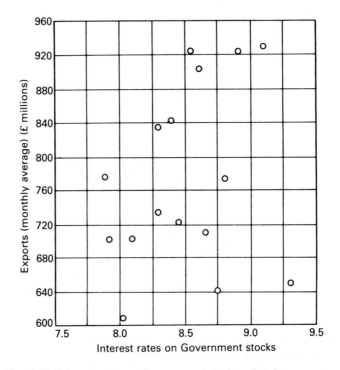

　　　　　　　　Fig. 6.12 Interest rates on Government stocks related to exports

Use and interpretation

The methods described in this chapter are extremely popular ways of presenting statistics without technical details. Examples may be found every day in newspapers, magazines, posters, bank reviews, and on television. Good, simple examples are popular with the person in the street because they are readily understood at a glance (they also often appeal to the statistician because he can express some of his artistic ability!).

Simplicity, however, must not be made an excuse for inaccuracy, and the desire to make pictures and symbols more attractive can lead to error as well as confusion.

Bar diagrams are the clearest kind of diagram to understand, as well as the easiest to draw. It is difficult to mislead the reader if accuracy in the width and height of bars is preserved. Bars on a diagram may be drawn touching each other or spaced out for clarity as in Fig. 6.3.

Pie diagrams are more difficult to draw, requiring compasses and a protractor. They do not give the reader a very clear idea of the proportions involved because the eye can measure lengths of bar much more easily and accurately than slices of pie. Again, to compare two or more pies of different sizes is difficult. For example, the reader might imagine that the sizes of the pies are proportional to their diameters, whereas it is the *areas* of the circles (squares of the radii) which are really important—but which are difficult to compare.

Pictograms can be misleading, especially if drawn in a haphazard manner.

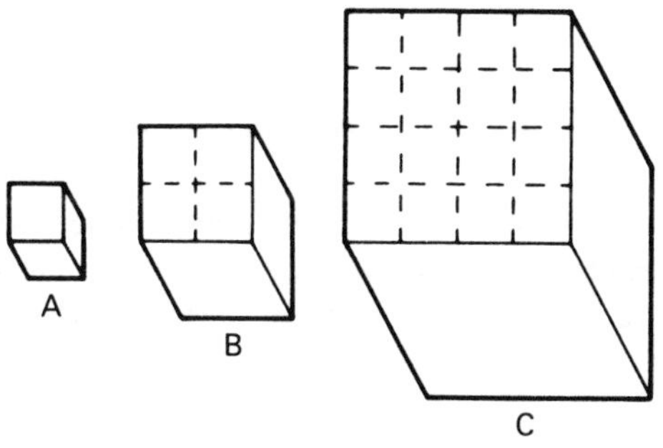

Fig. 6.13

Imagine that the cubes, A, B, and C, in Fig. 6.13 represent crates of exports. Suppose that we wish to show that in three years exports had doubled by the second year, and doubled again by the third year. That is, the size of exports (cube C) in the third year is four times the size of exports (cube A) in the first year.

Obviously, the symbols drawn in fig. 6.13 are wrong. Although the *length of a side* of the cube A has been increased four times in cube C, the *volume* of cube C is far larger than it should be.

The proper ratio of the exports should be $1:2:4$, but it is, in fact, $1^3:2^3:4^3 = 1:8:64$, if we judge by the symbols!

Of course, the same kind of error could be made in *two-dimensional* symbols. What would be the size of error if A, B, and C were *squares* instead of *cubes*?

The lesson here is that you should use only one symbol for each unit represented, and multiply the symbols as you represent multiples of the unit.

Cartograms

Cartograms are clear, attractive, easy to understand and free from misleading details, provided that a simple and accurate key is given.

Period change charts

Period change charts are often used as additional illustrations, or in cases where the absolute figures are quoted in the text of an article. This is perfectly valid. The danger is that the casual reader might interpret such charts too literally. They give no indication of the changes within the two dates where such changes may have varied greatly in positive and negative directions. The charts might mislead in that no indication of the size of items at the base date is given. Thus if Table 6.2 below was drawn on a period change chart, the change line for imports would be higher on the chart and steeper than that for National Income, even though the latter's absolute increase is more than double the former's.

Table 6.2

Item	Year 1	Year 10	Absolute increase	Percentage increase
National Income (£m)	27 191	52 404	25 213	93
Imports (£m)	4984	11 172	6188	124

Percentage changes in small items often appear huge when small absolute changes have taken place. These charts are best used when the items shown are of the same group (e.g. items within a total), and when the absolute sizes of the items are not too dissimilar.

Scatter diagrams

Though useful, scatter diagrams are a little too technical for the average reader to understand.

The point here is that, in an example such as in Fig. 6.9, the profit and turnover figures are simply compared, i.e. related. The worst mistake in interpreting this diagram is to imagine that one series of figures *causes* the other. A shop could have a low turnover with a high profit, or vice versa. In our example, a high profit happens to go with a large turnover. But good salesmanship might be the cause of a high turnover, and a good position in the centre of town might be the cause of high profits. One could hardly say that a good position in the centre of town causes good salesmanship. Thus, the two series of figures illustrated on the scatter diagram might have completely different causes to account for each.

It is natural to suppose that a scatter diagram shows how one series influences another, but here lies the gravest error. If one series might influence another (see Fig. 6.10), we may show it on a scatter diagram, but we must explain the connection in a way other than by graphs.

Statisticians often amuse themselves by finding examples of nonsense correlation, i.e. where the apparent cause of the relationship is not a true one. For example, there is a high degree of positive correlation between the number of television licences and the number of admissions to mental hospitals. Can we draw the conclusion that watching television drives people insane? We may be sorely tempted! The more likely explanation is that they are both related to some common factor, e.g. the increase in the size of the population, or the increasing industrialization of society.

Note. In all the examples of scatter diagrams which have been given, the number of pairs of items has been small—in most cases only 10. This is for purposes of illustration, and is also a common practice in examination questions. But whenever a real scatter diagram is constructed, the number of items should be considerably larger so that at least a hundred points appear on the diagram. This is because we are taking a sample, and if it is to reflect the behaviour of the whole group of items, then the sample should be fairly large.

Charts by computer

The quality of computer graphics from statistics packages varies greatly. Some packages are not very good at graphics in general; some may produce good screen graphics with the aid of special 'add-on' modules. How easy it is to print graphics on paper will also vary.

The rest of this chapter continues the introduction to MINITAB. Spreadsheet users should read and practise the computing sections at the end of Chapter 9 (Averages) and Chapter 18 (Further graphical applications) for spreadsheet graphics.

From data entry to chart

We are going to use the MINITAB statistical package again. We wish to produce a plot of the data represented by Fig. 6.1, the monthly sales of XY Ltd. First, we must input the data. Mainframe MINITAB users do not have the 'data entry' mode of the PC version of the package—the mode where we see the worksheet on the screen as we key in data. Mainframe users therefore have to enter the data into the computer using other software (the *editor*), or they can enter data from the keyboard to the MINITAB worksheet from within the MINITAB software. Within MINITAB we use the SET command (for inputting data one column at a time) or the READ command (for more than one column at a time). PC users also have these two options.

As always, each line of command is terminated by pressing the ⟨**RETURN**⟩ key. We first summon the package by typing:

At the 'MTB >' prompt we use the SET command to introduce the 12 monthly sales figures into column C1 of the (invisible) worksheet:

MTB > SET C1

The package responds by changing the prompt to 'DATA>'. We simply key in the data. Each month's sales figure must be separated by a space or by a comma. Imagine that the sales figure for December is missing. In MINITAB a missing data item is represented by the '⋆' symbol (without quote marks). Our data entry after the DATA> prompt therefore becomes:

```
DATA> 18000 21000 23000 30000 28000 36000
DATA> 20000 26000 30000 21000 20000 ⋆
DATA> END
MTB >
```

Notice that we can press ⟨**RETURN**⟩ anywhere in data entry, and continue on the next line. The '£' sign is omitted. Once the data are all keyed in, we type END after the DATA> prompt to let the package know that we are ready to key other commands. The package answers by returning to the MTB > prompt.

At this point the sales manager rushes in bearing the figure for December's sales, £10 000! Before we go any further we can therefore correct the worksheet, overtyping the '⋆' value. PC users can do this by summoning the data entry mode with the ⟨**Esc**⟩ key, and then overtyping the entry in row 12 of C1 directly (see Chapter 5). Mainframe users will need to use the familiar LET command, but this time for a data correction purpose:

MTB > LET C1(12)=10000

Here the LET command instructs the package to accept 10000 as the new value for C1(12), i.e. row 12 of column C1.

As explained in the previous chapter we can reassure ourselves that all is well with the data by typing INFO (to get a summary of the worksheet) or PRINT C1 (to see the 12 data entries). Assuming that all is well we move towards producing a plot of the data. Since we have time series data (i.e. over 12 months), an appropriate tool here is the TSPLOT command ('Time Series Plot'). Here is the exact form of the command we type after the MTB > prompt:

MTB > TSPLOT 12 C1

We have told the package that the time series cycle is 12, i.e. 12 months in a year. This detail helps MINITAB with labelling the output:

In this plot (Fig. 6.14) the months are labelled from 1 to 9 (January to September inclusive), then 0, A, B (October, November, December). The height of the symbol on the '£' (vertical) axis shows the level of sales achieved.

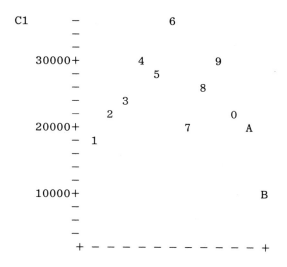

Fig. 6.14 Profits and turnover of 10 grocers' shops

Having produced this one, we now want to produce a plot of the grocers' shops data given in Table 6.1 earlier in this chapter. However, our worksheet currently contains the XY Ltd sales figures. How can we clear the worksheet without exiting from MINITAB entirely with the STOP command? Deleting the worksheet without leaving MINITAB is simply done by typing the command RESTART after the MTB > prompt:

MTB > RESTART

However, we should think carefully before typing this! The original data in the worksheet will be lost unless previously saved! Here, the loss of so little data is of no importance. But having keyed in bigger datasets we shall want to save data permanently. How to save data and to produce more charts and diagrams are dealt with in the following sections.

Having typed RESTART we wish now to enter the grocers' shops data (see above) and reproduce the scattergram in Fig. 6.9. For each of the 10 grocers we have two variables (i.e. two columns of data)—average weekly profit and average weekly turnover. We enter two or more columns of data into the MINITAB worksheet with the READ command. We specify which columns we want to fill; we type in the data on seeing the DATA> prompt, and we end the data entry with the END command:

More plots and commands

MTB > READ C1–C2
DATA> S25 2125
DATA> S29 2170

—and so forth for all the grocers until—

```
DATA> S74 2400
DATA> END
MTB >
```

A MINITAB scattergram is then easily produced with the PLOT command:

```
MTB > PLOT C1 C2
```

You should try all of this for yourself. Remember to follow each command with the (**RETURN**) key! If you would like to produce a scattergram that is rather better finished, try the following version of PLOT followed by some subcommands:

```
MTB > PLOT C1 C2;
 SUBC> TITLE 'Profit and turnover of 10 grocers shops';
 SUBC> YLABEL 'Av Weekly Profit £';
 SUBC> XLABEL 'Av Weekly Turnover £'
```

Notice that in the expanded version of the PLOT command above, the first line ends with a semicolon. This tells the package to expect subcommands before the scattergram can be produced. The package replies by changing the prompt to 'SUBC>'. We key the first subcommand to give our scattergram a title. This subcommand, too, ends with a semicolon. Two further subcommands (YLABEL, XLABEL) follow, giving labels to the C1 and C2 axes, respectively. The text for both title and labels should be typed within single quotation marks. The last subcommand, XLABEL, ends with a full stop, which tells the package that there are no more subcommands to follow, and that the PLOT should be executed. The MINITAB output is shown in Fig. 6.15. *Note:* Vertical lettering on *Y*-axis is unavoidable in MINITAB.

This PLOT will end our statistical analysis. However, before we leave MINITAB we wish to store the two columns of data in a file permanently kept on the computer's internal hard disk storage:

```
MTB > WRITE 'GROCERS' C1–C2
```

The WRITE command stores the two columns of data in C1 and C2 in a file called 'GROCERS'. The actual name must be typed within single quotation marks. MINITAB will know the file as GROCERS, but the operating system of the computer will have a slightly longer name for the file. This longer name is GROCERS.DAT (or GROCERS.DATA in some computer systems) to indicate that the file is a simple datafile.

If you are using a PC and wish to save the GROCERS file to an already *formatted* (i.e. electronically prepared for use) floppy disk in drive A, then you will have to type instead:

MTB > WRITE 'A:GROCERS' C1–C2

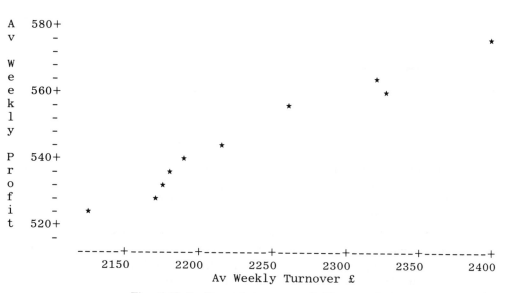

Fig. 6.15 Profits and turnover of 10 grocers' shops

There may be more complex instructions if you are saving data on your own personal area of a shared hard disk. Keep a careful note of the file name, and of where it is stored (hard disk or floppy).

You should not forget to complete your MINITAB session with:

MTB > STOP

7

Graphs

Method

The word 'graph' could well be associated in the student's mind with the word 'graphic', which means 'vivid', or 'springing to life'. This is exactly the kind of function which the graph should perform for the table from which it is drawn, because most people can grasp pictures more readily than figures.

This is not to say that graphs should be sensational. Graphs, or 'line charts' as they are often called, are really a diagrammatic type of representation, although one usually thinks of a graph as having extra technical features (i.e. statistical information) which would be out of place in popular diagram or symbol presentation. This may be a good place to say that the most accurate information that can come into the statistician's hands is the primary data, e.g. completed questionnaires and interviewers' answer sheets. From that point onwards, the data is subject to human errors of miscalculation and to the classification process (e.g. putting into class frequencies) in which some precision must be lost. After this 'boiling down' process it is then probably reshaped for the tabulation stage. It comes finally to the point where it is put into graphical terms and, as we shall see, it is further slightly distorted. It is up to the statistician to make the inevitable distortions at every stage as small as possible. It is certainly up to him to prevent and expose the tortured and dishonest examples of graphical work to be found in 'Use and interpretation' (pages 83–88).

Graphs generally
The graph is drawn on a grid (squared paper) to a certain scale. The basis of this grid is shown in Fig. 7.1.

The grid shows the four quadrants with zero, the origin, at the centre. In a graph, lines or curves are drawn on the grid to illustrate the relationship between two variables. As variables may be negative as well as positive, this is provided for by carrying the axes on to the left of zero on the horizontal axis, and below zero on the vertical axis. In practice, the student will not often meet with negative values and therefore there is usually no need to draw the negative arms of the axes.

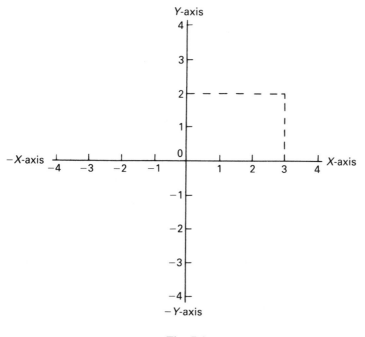

Fig. 7.1

DEPENDENT AND INDEPENDENT VARIABLES

The horizontal axis, known as the X-axis, is scaled in units of the independent variable. In a frequency distribution, the scale can be taken as the limits of the class intervals. In the graph of a time series, the independent variable on the X-axis is 'time'.

The vertical axis, known as the Y-axis, is scaled in units of the dependent variable. In a frequency distribution, the scale can be chosen according to the range of the class frequencies. In the graph of a time series, the dependent variable on the Y-axis is the value of the series at the regular intervals of time.

The independent variable is the one which is chosen to be stated in class intervals, and the class frequency is, therefore, the dependent variable. There is no certain way of choosing, say, the independent variable in most exercises concerning frequency distributions. For example, if we were to relate the two variables of output and cost in a firm, it might be impossible to decide which is dependent on which. Is the cost dependent on the output, or is the output dependent on the cost? From different points of view, each is true. There is no doubt, however, that students will be able to recognize the independent variable (and hence, the dependent variable) in their work, because these have usually been decided upon in the problems that will be presented to them.

67

PLOTTING THE CURVE

The curve joins the points of relationship of the two variables on the graph. (The word 'curve' is commonly used, even when a straight line is actually drawn in the graph.) The position of any point on the curve is decided by reference to the axes; e.g. in the basic grid in Fig. 7.1, a point has been plotted at:

$$+3(\text{on } X\text{-axis}) \text{ and } +2(\text{on } Y\text{-axis})$$

These are the 'bearings' or 'coordinates' of that point. The point of intersection is marked with + or, better still, by a ring. Either is better than × which does not reflect the vertical and horizontal nature of the axes. A set of points is thus built up, one for each pair of facts, and these are joined to form the curve.

DISCRETE AND CONTINUOUS VARIABLES

The curve may be a freehand smooth curve, or simply a joining of straight lines from point to point. Certain considerations govern which one is used, however.

If the dependent variable to be illustrated is discrete then, strictly speaking, the set of points should be joined by straight lines. If continuous, then the points should be joined by a smooth curve (i.e. one that is not sharply angled between one point and the next).

In practice, however, a certain latitude is sometimes allowed; e.g. a time series of National Income figures should not be plotted as a discrete series, although money is, strictly, a discrete variable. The steps between one point and the next, even if they represented £1 million, would appear as a smooth curve when the total National Income stands at its present figure of thousands of millions. On the other hand, yearly sales figures of a firm should be joined by straight lines, even though the totals involved might still be very large compared with the units involved. This is because we have no information about what is happening between one point and the next. There might actually be a drop in sales for certain months, and a continuous curve would suggest that there is a steady rise (or fall) in the figures, through time.

Finally, if more than one curve is drawn on a graph it is usual to distinguish the curves by labelling them separately and clearly, or by drawing the curves in different colours, or by drawing them in different ways e.g:

— — — — — — a broken line
— · — · — · — a dot–dash line
——————— an unbroken line
···················· a dotted line
xxxxxxxxxxxxxxx a hatched line
(000000000000000) a ring line

A key must be inserted on the graph unless the curves are separately labelled.

The histogram
68

This must not be confused with the historigram (graph of a time series—see Chapter 12). Students can avoid this confusion if they remember that *history* relates to *time*.

Let us suppose we are given the following frequency distribution in a table:

Table 7.1 Weekly bonuses in an industrial region

Weekly bonuses (£s)	Number of workers (thousands)
20–29.99	52
30–39.99	256
40–49.99	170
50–59.99	68
60–69.99	30
70 and over	24
TOTAL	600

To draw this as a histogram, the horizontal axis is scaled in class intervals and the vertical axis in 'Number of workers'. On each class interval is erected a rectangle the height of which is determined by the number of workers in the class in relation to the vertical scale. The last class in the frequency distribution is an example of an open-ended class. The usual method of treating such a class is to assume that it is of the same size as the classes immediately preceding it, i.e. in this case, 70–79.99, unless there is good reason to suppose otherwise.

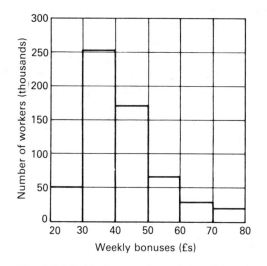

Fig. 7.2 Weekly bonuses in an industrial region

The total area of the histogram represents the total of the frequencies (600). The *area* of each rectangle represents the number of items falling into that particular class. This does not mean that the area *equals* the frequency, because we have scaled down the height and the width, according to our choice of units for the two axes. Nevertheless, the relative areas of the various rectangles are always strictly *proportional* to the class frequencies, whatever scale is chosen for the diagram. As all the class intervals in the distribution are equal, this is the same thing as saying that the height of each rectangle represents the number of items falling into each class in this particular case.

Quite often, in statistics, the rule that all class intervals should be equal is broken. In our example, the frequency distribution might have been written as follows:

Table 7.2 Weekly bonuses in an industrial region

Weekly bonuses (£s)	Number of workers (thousands)
20–24.99	12
25–29.99	40
30–39.99	256
40–49.99	170
50–59.99	68
60 and over	54
TOTAL	600

Here we see that the first two intervals are of £5 each, the last one (closed at 79) is £20, and the remainder are £10, as before.

The histogram may easily be adapted to deal with such unequal intervals. The width of each rectangle is determined by the class interval, but the heights must be adjusted to preserve the area relationships.

First, a 'standard' width of bar must be selected. This should be the class interval which occurs most often in the table. (The reason for this is to reduce the number of adjustments needed.) In the example this is the £10 group. Next, each frequency is divided by the actual class interval and multiplied by the standard. The resultant figures are the frequencies to be plotted, i.e. the heights of the various rectangles:

1. $\dfrac{12 \times 10}{5} = 24$
2. $\dfrac{40 \times 10}{5} = 80$

3. $\dfrac{256 \times 10}{10} = 256$
4. $\dfrac{170 \times 10}{10} = 170$

5. $\dfrac{68 \times 10}{10} = 68$
6. $\dfrac{54 \times 10}{20} = 27$

Hence, the first two bars are made twice as high (to compensate for the narrower width), the last bar is only half as high (because it is twice as wide as the rest), and the remaining bars are unaltered. In practice, no calculations would be needed for the groups of standard size.

A histogram would appear as follows:

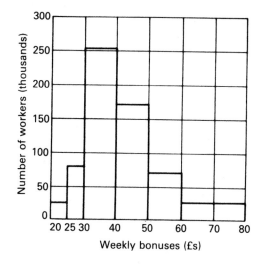

Fig. 7.3 Distribution of weekly bonuses in an industrial region

Unequal groups are often found in published statistics under the following circumstances:

1. When the original data falls naturally into certain groupings. For example, in an age distribution of factory workers, the first group would be '16 and under 21' since this would include the apprentices and so on. Thereafter, '21 and under 30', '30 and under 40', and so on, until we get '50 and under 65', and finally '65 and over', to include those of pensionable age.
2. When the items within a group are not spread evenly throughout it. This is the case in the example given. The first class interval of £10 has only 12 workers under £25, while there are 40 workers earning over £25. Splitting this group into two £25 groups reveals that the upper portion is heavily weighted with items.
3. When certain classes contain few or no items, it is better to combine them, in order to get a larger frequency.

The frequency polygon

This can be constructed from the histogram by joining the midpoints of the top of each rectangle. This is done by straight lines. The ends of the diagram (i.e. the two last points) may be left, or alternatively they can be joined to the base line at the centres of the adjoining class intervals. For example, the 80–90 group has no items; the centre point of the (imaginary) rectangle is therefore on the base line. The following diagram (Fig. 7.4) shows both methods, but in practice one or the other should be used.

If the question merely calls for a frequency polygon, there is no need to draw the histogram at all. We merely locate each point by reference to the given

71

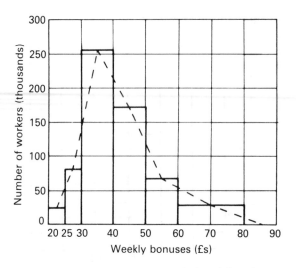

Fig. 7.4 Weekly bonuses in an industrial region

frequency (on the vertical scale), and the midpoint of the class interval (on the horizontal scale).

The frequency curve

Neither the histogram nor the frequency polygon gives a very accurate picture of a frequency distribution. This histogram suggests that frequencies are the same throughout the class interval—when they probably are not. The frequency polygon suggests that sharp, angular differences occur between the midpoints of the class intervals—when this also is probably untrue. However, they may give us a fair *idea* of the distribution, and, as long as only this is required, they will be useful.

A more accurate picture of the distribution would emerge if we could make two adjustments:

1. If we could draw a rectangle (say, in the histogram) for each individual item in the distribution or, at least, subdivide the class intervals, so as to make a greater number of smaller class intervals.
2. If we could increase the number of items in the distribution (if this were possible).

The histogram thus produced would have an outline, not of solid 'blocks', but one with much smaller 'steps'. It might appear like Fig. 7.5 on page 73.

This would give us a far more accurate histogram and we could derive from it a far more accurate frequency polygon. The smaller we make the class intervals and the greater the number of items we are able to include, the smoother the outline becomes.

The frequency curve would be the result if we could carry this process far enough to be able to draw a smooth, freehand curve, as in Fig. 7.6.

It is usually considered not possible to draw such a curve unless there are at least 1000 items in the distribution. This is because we need about 100 groups or

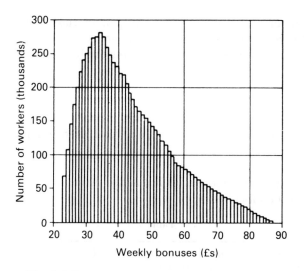

Fig. 7.5 Weekly bonuses in an industrial region

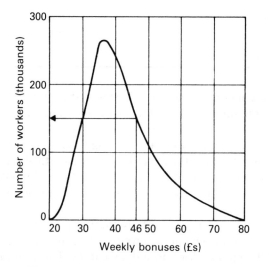

Fig. 7.6 Frequency curve from Table 7.1

rectangles in order to ensure small steps, and we also need a reasonable number of items in each group.

It is permissible to smooth a histogram or a frequency polygon if we believe, for example, that a *sample* distribution from which we constructed our diagrams is representative of the whole population, were it possible to tabulate the latter.

The frequency curve is a convenient way of drawing distributions in order to make comparisons easier, but two points should be noted about this curve:

73

1. The area under the curve between any pair of class limits represents the number of items occurring between those limits.
2. To draw a curve indicates that a true reading can be taken at any point on the curve.

For example, suppose we wish to estimate the number of workers earning bonuses of, say, £46. We draw a perpendicular line from £46 on the horizontal axis, to the point where it cuts the curve, and read off the number of workers on the vertical scale (see Fig. 7.6).

The cumulative frequency curve

To plot this curve, the frequencies are added cumulatively, i.e. in successive additions, as in the table below:

Table 7.3

Weekly bonuses (£s)	Number of workers (thousands)	'Cum' less (thousands)	'Cum' more (thousands)
20–29.99	52	52	600
30–39.99	256	308	548
40–49.99	170	478	292
50–59.99	68	546	122
60–69.99	30	576	54
70 and over	24	600	24
TOTAL	600		

The 'cum' less column shows, at any stage, the total number of workers earning *less* than the upper limit of that particular class interval. For example, 478(000) workers earn bonuses of less than £49.99 (or £50).

The 'cum' more column shows at any stage the total number of workers earning bonuses of more than the lower limit of the particular class interval. For example, 54(000) workers are to be found above the £60 point.

The two cumulative frequency curves are drawn on the graphs shown in Figs. 7.7 and 7.8.

The student should note most carefully that, in plotting the frequencies, the point must be placed at the *end* of the group interval for 'less than' curves, and at the *beginning* of the interval for 'more than' curves. This is because we have not reached our total of, say, 308 until we have counted all the workers in the first two classes, i.e. until we reach a wage of £39.99 (or £40).

Straight lines are conventionally used to join the points in these graphs because the distribution of items is assumed to be even. These curves are often known as 'ogives', a term used in architecture to describe a similarly outlined 'S' shape.

In the construction of these curves, no difficulty is caused by unequal classes; we merely join the unequally spaced points.

If two distributions are plotted on the same cumulative graph in order to compare them, the data should be reduced to the same scale, i.e. to percentages.

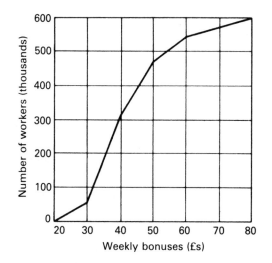

Fig. 7.7 Cumulative 'less than' frequency curve of weekly bonuses

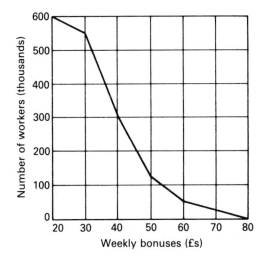

Fig. 7.8 Cumulative 'more than' frequency curve of weekly bonuses

The 'less than' curve is the one most frequently used and, if asked simply to draw an ogive, the student should draw this one. Further features of these graphs are discussed in Chapter 10.

Such curves are drawn on a ratio (or logarithmic) scale graph and they are used to show *relative* changes in data. In all the previous chapters we were concerned with absolute (often called arithmetic or natural) changes, i.e. changes of actual amount.

**The ratio
curve**

75

The student's work would probably be concerned with examples of the ratio curve mostly in connection with the graphs of time series. The graph below is a ratio graph of a time series and, as can be seen, only one axis (the *Y*-axis) is measured in a ratio scale. Therefore, this graph is known as a *semi-logarithmic graph*. It is not usual for the student at this level to meet examples of a full logarithmic graph, i.e. where *both* axes are scaled logarithmically.

The most important point in a ratio curve is not its position on the graph, but the *degree of slope* of the curve itself. Whereas the absolute curve measures the magnitude at any point, the ratio curve measures, at any point, the *percentage change from the last point*. Therefore any two equal distances measured on the logarithm axis will show equal percentage changes.

Suppose that, during a five-year period, a firm's sales were as follows:

Table 7.4

Year	1	2	3	4	5
Sales (£ hundreds)	2000	4000	6000	8000	10 000

On the absolute graph and the ratio graph these would appear thus:

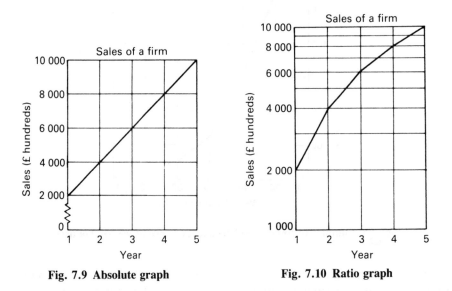

Fig. 7.9 Absolute graph   Fig. 7.10 Ratio graph

On the absolute graph, the distance from 2000 to 4000 is, of course, exactly the same as the distance from 4000 to 6000; but on the ratio graph, the distance from 4000 to 6000 is much smaller than the distance from 2000 to 4000. This is because the ratio graph measures the percentage change on the last figure, i.e:

2000 to 4000 = an increase of 100 per cent
whereas 4000 to 6000 = an increase of 50 per cent

Therefore, in the case of the absolute graph, the simple magnitude of the series appears as a straight line. On the ratio graph, *using the same figures*, the curve shows a falling percentage increase of each year on the year before. To put the matter another way, the distance on the *Y*-axis, on the ratio graph, from 2000 to 4000, is exactly the same as the distance from 4000 to 8000 because, although the absolute increases are 2000 and 4000 respectively, the percentage increases are exactly the same, i.e. 100 per cent.

There are two points to notice about the ratio graph:

1. There is no zero base line on the graph (because the logarithm of 0 is minus infinity, which is impossible to show). Similarly, negative values cannot be plotted.
2. The time axis (*X*-axis) is scaled in ordinary absolute measure.

TWO METHODS OF DRAWING A RATIO GRAPH

1. Obtain semi-logarithmic graph paper. Mark off the time axis with an absolute scale. Mark off the axis of the variable according to the range of the actual variable. Plot the actual figures of the variable and join the points.
2. If you have no semi-logarithmic graph paper, find the logarithms of the values you wish to plot. Using graph paper with ordinary absolute scale grid, mark off the time axis with an absolute scale. Measure the scale of the *Y*-axis using an appropriate range of the logarithms you have found:

Table 7.5

Year	Actual value	Logarithm
1	2000	3.3010
2	4000	3.6021
3	6000	3.7782
4	8000	3.9031
5	10 000	4.0000

Pencil in the logarithm figures (shown on left of Fig. 7.11) and plot the logarithm values. Erase the pencilled figures and mark in the actual values on the *Y*-axis.

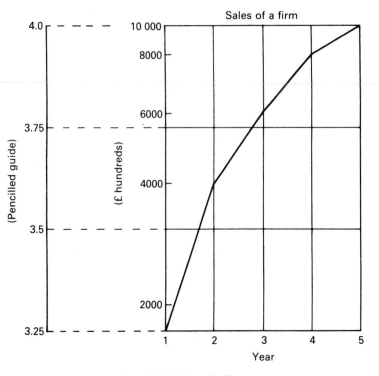

Fig. 7.11 Sales of a firm

A final example of a ratio scale graph is given below. Two curves have been plotted from the following data relating to a firm (see Fig. 7.12):

Table 7.6

Year	Number of operatives employed	Bank loans (£ thousands)
1	734	650
2	622	1100
3	510	2500
4	827	11 000
5	1050	9050
6	1804	5000
7	2113	5000
8	3241	6000

This is an example showing how two series of completely different types, i.e. one in a monetary unit, the other in human units, and of quite differing magnitudes, may be plotted on the same graph with the same scale. This would be difficult on an absolute graph, because two different scales would be

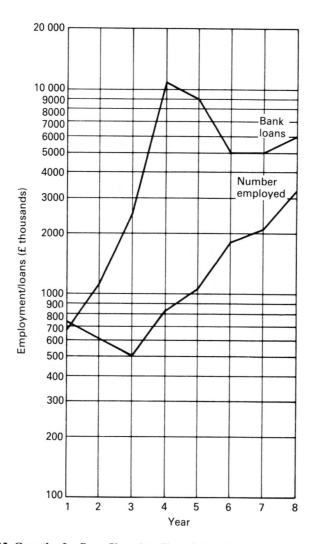

Fig. 7.12 Growth of a firm, Year 1 to Year 8. Employment and bank credit

required; the curves would be more widely separated (thus making comparison difficult); and there would be no common basis for comparison (here there is the basis of the rate of change of each series—*rate of change* being the common factor).

The two ratio curves in the example have been plotted on three-cycle semi-logarithm graph paper. That is to say, there are three blocks or 'cycles' on the logarithmic axis scaled to measure the series. Because the total range of all the actual values, on both tables, is 510 to 11 000, it was decided to use all three cycles—the lower cycle being measured in hundreds (100 to 1000), the middle cycle being measured in thousands (1000 to 10 000), and the upper cycle measured in ten thousands (10 000 and upwards). The three cycles, measured in this way, are quite sufficient to include all the actual values. In a similar manner, the student must adapt the ratio grid to the figures to be dealt with.

For example, the first cycle may be in units, the next in tens, and the third in hundreds. The essential point is that the scale of each cycle must be 10 times that of the previous one.

Note. Each cycle or block is subdivided into 9 main sections (not 10, as in ordinary graph paper).

Time series

The analysis of a time series, with graphical examples, is dealt with in Chapter 12. It is sufficient to say here that if we wish to compare two or more curves on one time series graph we can do this in two basic ways:

1. All curves which are of the *same nature and in the same units* (e.g. bank advances, and bank investments) can be plotted using one axis scaled in those units. The curves will then be directly comparable.
2. Two curves which are *different in nature or are in different units*, e.g. exports and savings (different nature), or unemployment and bank deposits (different nature and different units), can be plotted using the left-hand vertical axis for one type of unit, and the right-hand vertical axis for the other. In these cases, although the *rise and fall* of the curves may be compared, the curves as a whole are not directly comparable (see Fig. 7.12).

As has been mentioned previously, the *X*-axis is always chosen to show the time scale, and the *Y*-axis to show the variable. Confusion often arises when plotting from the time axis. If the variable values are known to be mid-year, or mid-monthly figures, for instance, then the points should be plotted *between* the yearly or monthly intervals along the *X*-axis. In practice, say, for mid-year figures, points are usually plotted *on* the yearly limits, and a special note is included on the graph to allow the reader to make the appropriate mental correction. For example, 'Figures relate to total deposits on 30 June each year', or 'Yearly figures of total deposits are averages of the monthly totals'.

Band curve charts

Band curve charts are sometimes known as 'layer' graphs. This form of time series graph is often used where totals over a period can be broken up into constituent parts. For example: *bank assets*—cash, bills, advances, investments; *textiles*—wool, cotton, jute, man-made fibres.

A typical graph is shown in Fig. 7.13.

Note. The additional vertical scale at the right-hand side, and also the patterning of the bands, are to help the reader.

The series, of course, must each be subtotals of a total group in the same units.

In this particular case, the curves are plotted directly from actual figures, but, when this is not convenient, secondary statistics can be derived, e.g. by reducing the actual values to percentages of the total (100 per cent).

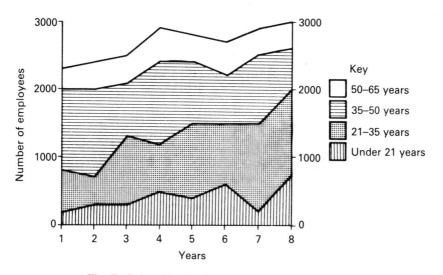

Fig. 7.13 Age distribution of employees, XYZ Ltd

1. The title must be clearly stated on the graph, and it should be written on last so as not to interfere with the curves on the graph. **Rules for graphical work**
2. All axes should be scaled and labelled with the variable they represent, and the units of the variable should be shown where necessary.
3. All lettering on graphs should follow the normal horizontal pattern except in the case of vertical axes (right and left) when lettering should be done as at (a) and (b), but not (c):

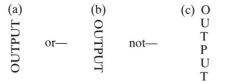

4. Sources and footnotes, and a key, should be provided when necessary in a clear space on the graph.
5. In a natural (arithmetic or absolute) scale graph the zero line must be shown.
6. The graph should not be loaded with many curves, especially if the lines of the curves tend to run closely together (in the case of histograms it is inadvisable to draw two on one graph).
7. Plan the graph well, taking careful note of the range of each variable, and draw the graph to fill the available paper. Graphs which are crowded into a corner of the paper are usually the result of bad initial planning.
8. If ratio (logarithmic) scales are used, this should be quite plainly shown on the graph, unless, of course, the logarithmic grid is shown.

9. If percentage scales are used, some indication of the absolute values to which they refer should be made on the graph.
10. If the application of rule 5. would result in much blank paper, and a compression of the vertical scale (see Fig. 7.14), this may be overcome in two ways:

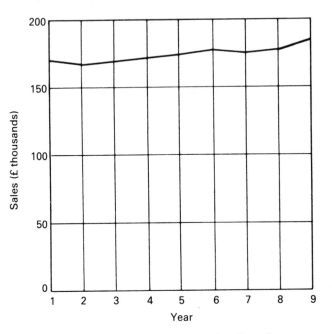

Fig. 7.14 Sales of a firm, Year 1 to Year 9

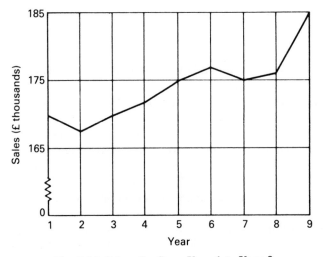

Fig. 7.15 Sales of a firm, Year 1 to Year 9

(a) The vertical scale may show a 'pleated' or 'broken' portion, to indicate that some of the (empty) space has been compressed (see Fig. 7.15).
(b) The vertical scale may be broken by two jagged lines running across the diagram to indicate that a portion of the (empty) space has been omitted (see Fig. 7.16).

Note. In each case, a much more open vertical space is possible which reveals any movement in the graph more clearly.

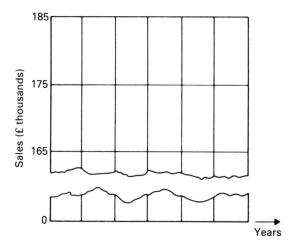

Fig. 7.16 Sales of a firm, Year 1 to Year 9

Use and interpretation

Ratio scales:	*Absolute scales*:	**Comparison of ratio and absolute scales**
1. Show the rate of change of a variable.	Show the actual size of the change in a variable.	
2. No zero base line.	Zero base line included.	
3. Negative quantities cannot be shown.	Negative quantities can be shown.	
4. On the ratio axis, equal distances represent equal percentage changes.	On the absolute axis, equal distances represent equal actual values.	
5. Where high or low extreme values are met with (e.g. 300 to 1 million) these can be shown easily and clearly.	Extreme values are often difficult to show clearly on a fairly small graph. For example, if 300 is represented by one small square, 1 million would need more than 330 large squares.	

Ratio scales:

6. Ratio graphs can be used to compare directly changes in variables which are not of the same nature or units. The ratio scale reduces them to a common base.

Absolute scales:

Absolute graphs can be used to compare, only indirectly, changes in variables which are not of the same nature or of the same units.

In addition, ratio scales can be useful when the variable fluctuates violently, e.g. during strikes, war, etc. Such fluctuations are dampened by the use of ratio scales, and they do not dominate the graph.

Ratio scales are particularly appropriate for series where the current figures depend directly on past figures for their magnitude, e.g. future population and past population, birth-rate and population.

Examples of bad graphs

Suppose that a firm markets a drink which will 'release one's hidden stores of energy' and that it begins a sales campaign, advertising through various media, newspapers, magazines, posters and television. Suppose, further, that it employs a statistician who is not unduly troubled by the standards of his profession. The firm might publish the following graphs in its advertising matter. The name of the product we will call *Zippy*. Sales are taken over a five-year period.

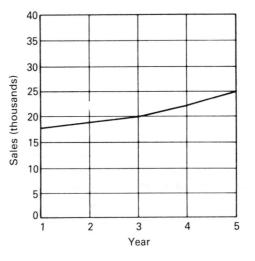

Fig. 7.17 Sales of Zippy

There appears to be nothing wrong with the graph shown in Fig. 7.17, although actually the reason for the steeper increase in sales from Year 3 was partly because of a heavy cut in the price of *Zippy*—and were we to graph the sales curve in *value* terms (instead of units sold), the curve would be much less steep. Although the graph tells nothing of this, we cannot fairly complain.

However, it appears that the graph is not striking enough, and the manager suggests to the statistician that the graph might look a little better in the form of Fig. 7.18.

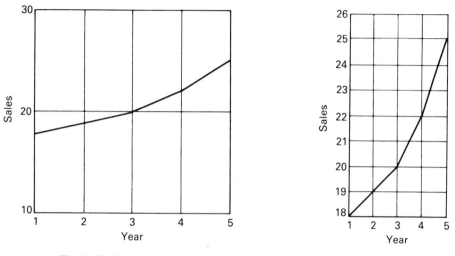

Fig. 7.18 Sales of Zippy   **Fig. 7.19 Sales of Zippy**

Here we see that the extreme ends of the vertical scale have been cut off, and this gives the impression of a rather steeper rise. It is contrary to the rules of good graph-making, of course, to begin above zero, unless the axis is 'broken', and even that is not necessary here. Units have also been missed off the vertical scale.

Not content with the previous distortion, the firm produces an even more sensational graph, as shown in Fig. 7.19.

The tremendous rise in sales indicated here is achieved simply by spreading out the narrow band of sales, on the vertical axis, within which actual sales took place, and crushing together the periods on the time axis. Sales appear to have risen from almost nothing to almost the limit of possibility. The student can see that the steepness of a curve can be arranged on any graph merely by altering the scales on each axis and cutting down the frame of the graph. This graph offends the rules even more than the second graph, because definite intent to deceive is more plainly apparent.

Having sinned twice, it seems to the firm worth little to forbear once more. So a further graph appears in the advertisements—this one calculated to have even greater emotional appeal (see Fig. 7.20).

The idea of the graph is to imply that *Zippy* will penetrate the 'tiredness barrier', which prevents 'energy release'; other drinks will not. It will be noticed that no axes are labelled either by title or unit, neither are the axes scaled. Presumably the horizontal axis is 'time', but the statistician is probably contemptuous of the kind of people who would be impressed by such a 'graph' anyhow and has not bothered to record minutes, hours, etc. The vertical axis apparently measures 'tiredness depth' in some mysterious way. As this is unmeasurable, it is pointless to include it on the graph anyway. The 'tiredness barrier' appears to be sheer nonsense.

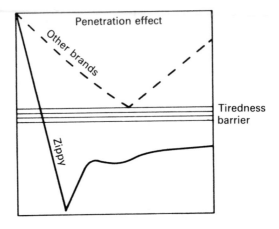

Fig. 7.20 The tiredness barrier

A further way of increasing sales is, of course, to invade your competitors' market. To compare the national sales of similar beverages with *Zippy*, the firm's statistician drew the graph shown in Fig. 7.21.

The statistician did not submit this to the manager for publication, however, for fear of receiving the answer that the graph lacked conviction. A few minutes more reflection turned possible defeat into victory and produced the surprising picture given in Fig. 7.22.

Fig. 7.21 Comparison of Zippy sales with other beverages

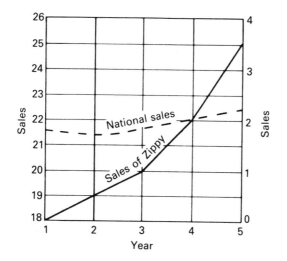

Fig. 7.22 Comparison of Zippy sales with other beverages

The effect in Fig. 7.22 is produced simply by placing two different vertical scales at each side of the graph and then manipulating them so as to produce the most favourable effect for one particular curve. The axes are scaled, but the too casual reader, it is hoped, will not notice the different scales. For instance, why should only one axis begin at zero? Note, also, the lack of units.

It appears in Fig. 7.23 that the promoters of *Zippy* have become obsessed by delusions of future grandeur. In this graph they have actually drawn a curve between only two plotted points, and these points range over a period of 18 years. On this pathetic basis they have projected into the future in respect of

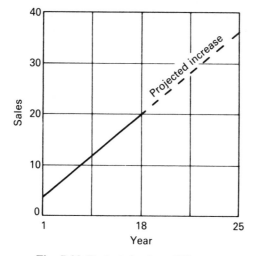

Fig. 7.23 Projected sales of Zippy

sales in a way that no honest statistician would ever dream of doing. The student will notice that Year 25, only seven years from Year 18, is marked at the same distance from Year 18 as is Year 1, where the difference is 18 years!

These examples do not, of course, exhaust the appetite for misrepresentation of firms like this imaginary one, and the student is urged to scan current literature for further examples. Unfortunately, these will not take long to find.

Graphs by computer

Other packages   Users of packages other than MINITAB should consult Chapters 10 and 11 (SPSS^x), Chapter 11 (SPSS/PC+) and Chapters 9 and 18 (Spreadsheets) for how to produce some simple graphs in other packages.

Recalling data in MINITAB with READ   The rest of this chapter looks at some more aspects of graphs by computer, and almost completes our rapid introduction to the MINITAB package. At the end of Chapter 6 we saved our data from 10 grocers' shops in a computer data file (kept on hard disk or floppy disk). The data was stored as the MINITAB data file 'GROCERS'. Returning to this file some days after it was created we first summon the MINITAB package and then load the stored data into the blank worksheet:

```
minitab

(–starting messages appear on the screen–)

MTB > READ 'GROCERS' C1–C2
```

Summoning a file from diskette, as opposed to hard disk, will involve also giving the drive name (e.g. 'A:GROCERS'). Exact details of how to recall files will vary by computer installation. The READ command reads the file (its name surrounded by single quotation marks) into the specified columns (here C1–C2) of the worksheet. The package responds by showing us the first four lines of data on the screen (the rest is present, but not shown on screen). When we see the MTB > prompt we can proceed with analysis.

MINITAB histograms   We want to produce a histogram of the data in column C2. This is easily done with the HIST command. The output appears immediately:

```
MTB > HIST C2

Histogram of C2  N = 10

Midpoint     Count
     150         3   ***
     200         3   ***
     250         1   *
     300         2   **
     350         0
     400         1   *
```

Notice how different the output is from the sort of histogram shown earlier in this chapter as Fig. 7.2. The MINITAB histogram is displayed on its side. The height of the bars is represented by the number of star (★) symbols. The number of observations in each class ('Count') is also displayed. The width of each class (the class interval) is 50 (£s). We are given the class midpoints rather than the class limits or boundaries.

It is possible to control the histogram display by adding subcommands to the HIST command. The subcommand INCREMENT specifies the class interval; START specifies the midpoint of the first class. Paying regard to the ';' and '.' symbols that indicate where subcommands are to start and finish, key the following version of the HIST command. You will first need to have read the GROCERS data into the worksheet:

```
MTB > HIST C2;
SUBC> INCREMENT 50;
SUBC> START 125.
```

Compare the output with that of our first attempt. The file GROCERS will remain stored but if there have been any alterations to the data, these alterations should be saved by re-saving the entire file with the WRITE command introduced in the last chapter. You should end your MINITAB session with STOP, or clear the worksheet with RESTART in order to continue with the next example.

Hard copy from MINITAB

This far, all our analysis work has been displayed on the screen, but then lost from view as it scrolls up and off the screen. Fairly soon we shall want to print *hard copy*, i.e. a paper record of the analysis.

PC users can send to the printer any screenful of interaction with the computer by holding down the ⟨**SHIFT**⟩ key and pressing the ⟨**PrtSc**⟩ key at the same time (⟨**SHIFT**⟩+⟨**PrtSc**⟩). This simply sends the contents of the present screen to the printer. This routine can be repeated at any time, either within or outside the MINITAB package.

Within MINITAB, users of all versions of the package can send a record of commands typed and output created with the PAPER command. After the PAPER command, everything that appears on the screen will also be sent to the printer. The PAPER command is turned off with NOPAPER, returning us to screen output only:

```
MTB > PAPER
MTB >                     Command typed here
MTB >                     Another command
                          Output begins
                          Output finishes
MTB > NOPAPER
```

However, both these methods of getting paper output are usually unsuitable in large offices and educational computing laboratories. In many locations a single printer is shared by many users via a *data switch*, a switch that allows the printer to be connected to only one computer or computer terminal at a time. Very often, a better solution is to put all our results into an electronically stored *output* file. This file is then sent to the printer at the end of the session, once we have exited from MINITAB with STOP. Do not confuse the *output file* with the *data file* which contains the basic observations from which the output is generated. The next example illustrates the use of an output file and the creation of a special data file called a *workfile*.

A layer graph with MINITAB

In this next exercise we produce a *layer graph* or *band curve graph* (see page 80). We call up MINITAB in the usual way. Using the READ command or the PC 'data editor' mode, we enter into the worksheet 10 rows of data giving processing/materials costs for a small manufacturing company. There are 10 rows (10 years) of data for three variables or columns (C1–C3)—steel, chemicals and energy. The display below is produced with PRINT C1–C3:

ROW	C1	C2	C3
1	6	16	33
2	6	20	33
3	4	15	39
4	3	14	32
5	5	15	34
6	8	21	38
7	10	24	39
8	10	28	41
9	11	30	44
10	13	30	47

For this session, everything that appears on the screen will be sent to storage in an output file. At the MINITAB prompt we type:

```
MTB > OUTFILE 'COSTS1'
```

The name 'COSTS1' is a file name of our own choosing. Notice that it is introduced to MINITAB within single quotation marks. If we are using a diskette in drive A to save files, then the command would look as follows:

```
MTB > OUTFILE 'A:COSTS1'
```

We have decided to give the three variables (or columns of data) names:

```
MTB > NAME C1 'STEEL'
MTB > NAME C2 'CHEMICAL'
MTB > NAME C3 'ENERGY'
```

Having named the variables thus, we want to produce a layer graph of the sort illustrated in Fig. 7.13. Notice that having named the variables we can use the names rather than the original column designations (C1, C2, C3) in the appropriate command. Here the appropriate command is MTSPLOT—Multiple Time Series Plot:

MYB > MTSPLOT 'STEEL' 'CHEMICAL' 'ENERGY'

The output that follows (Fig. 7.24) uses the same labels as the TSPLOT command (see Chapter 6) for the first variable ('STEEL')—i.e. 1 to 9, 0. The second variable ('CHEMICALS') is labelled by 'Y'. The final variable ('ENERGY') is marked by 'Z'.

```
                -                                         Z
         45.0+                                    Z
                -                              Z
STEEL           -             Z          Z   Z
                -
                -      Z   Z      Z   Z
         30.0+                                 Y   Y
                -                           Y
                -                      Y
                -         Y           Y
                -
         15.0+    Y       Y   Y   Y
                -                               9   0
                -                     6   7   8
                -      1   2          5
                -             3   4
          0.0+
                +-----+-----+-----+-----+-----+
                0     2     4     6     8     10
                Y = CHEMICAL  Z = ENERGY
```

Fig. 7.24

Now that our statistical analysis is complete we close COSTS1, the output file into which everything appearing on the screen has been placed for later printing on paper. We close COSTS1 with:

MTB > NOOUTFILE

We have not quite finished our MINITAB session. We wish to store away the three columns of data, plus the names given to the columns (and constants (e.g. 'K1', 'K2') if we had created any here). In Chapter 6 we stored away the GROCERS data with WRITE, and we recalled it in this chapter with READ. This pair of commands is inappropriate in the present case because we want to store variable names, like STEEL, as well as the three columns of data. The appropriate command to create this sort of workfile (i.e. data in columns, plus variable names, plus stored constants) is SAVE:

File management and hard copy

MTB > SAVE 'COSTSERS'

(We use the alternative 'A:COSTSERS' for diskette systems.) The package responds with the message:

```
Worksheet saved into file: COSTSERS.MTW
```

This tells us that the data plus names have been stored in the file 'COSTSERS' (we have chosen this name to indicate data on COST time SERieS). Although MINITAB recognizes the file name COSTSERS, the operating system of your computer outside MINITAB will only recognize the file by the longer name COSTSERS.MTW. The suffix '.MTW' (or '.MTW' on some systems) indicates that COSTSERS is a MINITAB workfile.

Finally within MINITAB we type the familiar STOP command to exit from the package back into the general operating system of the computer. Should we wish later to re-use the data and names stored in COSTSERS we would first summon MINITAB. At the MTB > prompt we would key:

```
MTB > RETRIEVE 'COSTSERS'
```

Notice from the above command that single quotation marks are used, and that we do not specify the columns. To recall the file from diskette we should have to specify the fuller 'path name' such as 'A:COSTSERS'.

Now that we are outside MINITAB we can send the output file, COSTS1, to the printer. The exact form of the command for this will vary between computer installations. There is one complication, however, that will be common to virtually all systems. Whereas MINITAB knows the output file created here as COSTS1, the operating system will know it as 'COSTS1.LIS,' a slightly longer name. The suffix '.LIS' (or '.LIST' in some systems) indicates that this is a LISTing file generated by MINITAB.

To print the output file COSTS1 on paper from a PC-style computer you will need to key a command followed by (**RETURN**). Outside the MINITAB package single quotation marks are not needed:

```
        PRINT COSTS1.LIS       (from storage on hard disk)
or
        PRINT A:COSTS1.LIS     (from storage on diskette)
```

At first it may seem confusing to work with several different file types, and so some instant revision of file handling is given below.

In the above exercise we created a file to collect all the screen displays (principally the layer graph) before printing outside the statistical package. This *output file* was created with OUTFILE, and closed with NOOUTFILE. Its name, chosen by ourselves, was COSTS1, but the operating system knows it by the longer name COSTS1.LIS or COSTS1.LIST.

In this example we also wanted to store the data that gave rise to the analysis. The data, plus variable names, were stored in the *workfile* COSTSERS. This was created with SAVE. The file can be recalled to the worksheet in a later

MINITAB session with the RETRIEVE command. The file name chosen by ourselves was COSTSERS, but the operating system outside MINITAB calls the file COSTSERS.MTW or COSTSERS.MTWK.

For all types of file, MINITAB requires file names to be typed within single quotation marks. The operating systems of PC-type computers do not use quotation marks around file names. Other computers may vary in their procedures.

8

Accuracy, errors and percentages

Method

Most numbers found in published statistics are not strictly accurate (i.e. they have been 'rounded'), for reasons to be discussed later.

Rounding numbers

This is a method of expressing the approximate value of data. We are usually asked to express a number to the nearest hundred, or some similar unit, but sometimes we work to so many 'significant figures'.

An example might be the value 132 854.54 kg.

1. If we wished to express this to the nearest hundred we should first determine how many hundreds are contained in the number—1328 $\frac{54.54}{100}$ hundredths of kg. If the remainder, or fractional part of a hundred is *exactly one half or more*, then we increase the total of whole hundreds by one. The answer in this case would be 132 900.

   This rounding may be done for other units as follows:

   132 854.54 kg to the nearest whole number = 132 855
   132 854.54 kg to the nearest ten   = 132 850
   132 854.54 kg to the nearest hundred   = 132 900
   132 854.54 kg to the nearest thousand   = 133 000
   132 854.54 kg to the nearest ten thousand = 130 000

   If the method of rounding is not stated, we can determine it by the number of noughts in the estimate, for example, when there are three noughts, it is in thousands, and so on. Note also that 132 900, to the nearest hundred, means within the range 132 950 to 132 850. So far as we know, the true figure could lie anywhere between 50 kg either side of the approximation. We usually write this as 132 900 ± 50 kg (i.e. half the number of noughts).

   Similarly, when we see at the head of a table the unit stated as 'kg (thousands)', we know that each figure in the table is subject to an error of ±500 kg.

2. Another form of rounding occurs when we are asked to express a value to a certain number of significant figures.

A significant figure is usually considered to be any number from 1 to 9 (inclusive). But when zero occurs between two significant figures, then zero is regarded as a significant figure.

For example, express 34.801, 42 367, and 6015.2 to three significant figures.

$$34.801 = 34.8$$
$$42\ 367 = 42\ 400$$
$$6015.2 = 6020$$

3. One might be asked to find an answer 'correct to two decimal places'. In this case, the student should work out the answer to three decimal places:

$$474.328$$

and then round it to two decimal places. In other words, work out one more than is asked for, and then round to the correct number:

$$474.33$$

We may summarize the degrees of accuracy which one finds stated formally, for the figure 5672.835 grams: **Degrees of accuracy**

1. To the nearest hundred = 5700 grams.
2. To three significant figures = 5670 grams.
3. Correct to one decimal place = 5672.8 grams.

POSSIBLE ERRORS **Errors**

Where the actual or true value is not known, it is usual to express the degree of possible error in two more ways:

4. 5500 ± 500 grams.
5. 5600 ± 3 per cent.

These two ways show the possible error either side of the approximate, or estimated, value: in 4. the possible range in which the true figure lies is 5000 grams to 6000 grams; in 5. the possible range in which the true figure lies is 5432 grams to 5768 grams.

ABSOLUTE ERRORS

The absolute error is the difference between the actual, or true, value and the approximate (rounded or estimated) value:

$$\begin{aligned} \text{Actual value} &= 3752.9 \text{ kg} \\ \text{Estimated value} &= 3800 \text{ kg} \\ \text{Absolute error} &= +47.1 \text{ kg} \end{aligned}$$

When an estimate is stated as 3800 ± 50, then the 50 is the absolute error, since the true value is not known.

RELATIVE ERRORS

The relative error is found by expressing the absolute error as a percentage of the actual, or true, value. (If the actual value is not known, the absolute error is expressed as a percentage of the estimated value—the difference here will not actually be large.) For example:

$$\begin{aligned} \text{Actual value} &= 3752.9 \text{ kg} \\ \text{Absolute error} &= +47.1 \text{ kg} \\ \text{Relative error} &\quad \frac{+47.1}{3752.9} \times 100 = +1.26 \text{ per cent} \end{aligned}$$

or

$$\begin{aligned} \text{Actual value} &= 3752.9 \text{ kg} \\ \text{Estimated value} &= 3800 \text{ kg} \\ \text{Absolute error} &= +47.1 \\ \text{Relative error} &\quad \frac{+47.1}{3800} \times 100 = +1.24 \text{ per cent} \end{aligned}$$

Conversely, a relative error can easily be turned into an absolute one:

$$3800 \text{ kg} \pm 3 \text{ per cent} = 3800 \text{ kg} \pm \left(\frac{3}{100} \times \frac{3800}{1} \right)$$

$$= 3800 \text{=kg} \pm 114 \text{ kg}$$

An absolute error is always in the same unit as the figure to which it refers, but a relative error is not in any unit—it is a pure percentage—the units cancel out in the division sum.

CUMULATIVE (BIASED) ERRORS

This kind of error is produced when the errors in a table of figures are all in the same direction (see the example in Table 8.1). This would mean, for instance, that the greater the number of items in the table to be added, the larger the resulting error. (In Table 8.1, the figures in the last two columns are taken to the highest 10 above the actual figures, and to the lowest 10 below the actual figures.)

These occur when the errors in a table of approximate figures tend to cancel each other out. Thus, when the figures in Table 8.1 were rounded to the nearest 10, the total came very close to the true addition of the actual figures. In this case, when the number of items in the table to be added is greater, the compensating error tends generally to diminish.

Table 8.1 Weekly earnings of male workers

Earnings in £s	True figure	Compensating (rounded)	Cumulative (highest 10)	Cumulative (lowest 10)
80 and under 90	41	40	50	40
90 and under 100	62	60	70	60
100 and under 110	87	90	90	80
110 and under 120	96	100	100	90
120 and under 130	32	30	40	30
130 and under 140	39	40	40	30
TOTALS	357	360	390	330
Absolute errors		+3	+33	−27
Relative errors		+0.8%	+8.5%	−8.2%

As can be seen, the compensating error, achieved by rounding, is far smaller than when the errors are cumulative.

1. *Addition*   When approximate values are added, the total error is the sum of the separate absolute errors in the values. **Laws of errors**
   e.g. 300 to the nearest 10, plus 400 in the nearest 100:

$$300 \pm 5$$
$$400 \pm 50$$
$$= 700 \pm 55$$

2. *Subtraction*   When approximate values are subtracted, the total error is the sum of the separate absolute errors in the values.
   e.g. 400 to the nearest 100, minus 300 to the nearest 10:

$$400 \pm 50$$
$$300 \pm 5$$
$$= 100 \pm 55$$

The error here is so large as to make the estimated answer useless. This is always a danger with subtraction, and in an actual business situation, the estimates must be made as accurately as possible, even if this takes longer.

3. *Multiplication* When approximate values are multiplied together, the total error is approximately the sum of the relative errors in the values.
e.g. 300 to the nearest 10, times 400 to the nearest 100:

$$\left(300 \pm \frac{5}{300} \times 100\right) \times \left(400 \pm \frac{50}{400} \times 100\right)$$

$$= (300 \pm 1.67 \text{ per cent}) \times (400 \pm 12.5 \text{ per cent})$$
$$= 120\ 000 \pm 14.17 \text{ per cent}$$

What we have done here, is to convert the absolute errors into relative errors, and then add them. A similar procedure is used in division.

4. *Division* When approximate values are used in division, the total error is approximately the sum of the relative errors in the values.
e.g. 400 to the nearest 100, divided by 300 to the nearest 10:

$$\left(400 \pm \frac{50}{400} \times 100\right) \div \left(300 \pm \frac{5}{300} \times 100\right)$$

$$= (400 \pm 12.5 \text{ per cent}) \div (300 \pm 1.67 \text{ per cent})$$
$$= 1.33 \pm 14.17 \text{ per cent}$$

The student will find it easy to apply these rules if the following points are remembered:

1. Use absolute errors for addition and subtraction. Use relative errors for multiplication and division.
2. Always add errors—*never* subtract them.

Percentages   Ratios and percentages (like the relative errors described above) tell us of the *rate* of change in given values. If we confuse them with *absolute values* we are likely to give our statistics a misleading turn and the reader a wrong impression.

Example:
In area A the number of deaths from cancer has risen by 400 per cent, while in area B, over the same period, the number of deaths from cancer has risen by only 29 per cent.

One may receive the impression that area A is somehow a much less healthy place. The true position is as follows:

Table 8.2 Deaths from cancer in two areas

	Population	Deaths from cancer (1)	Deaths from cancer (2)	Absolute increase	%
Area A	200	10	50	40	400
Area B	120 000	350	451	101	29

The absolute figures of increase for area B are far greater than those of area A. Also, area A may well be a small country place with a high proportion of old people in which we would expect cancer deaths to occur with a high (and highly variable) frequency. This may have been a bad year.

Misleading figures can produce even more misleading impressions and interpretations. In this case, we must either give the reader all the relevant figures, or we must somehow scale down the exaggerated percentage (400), to take account of the very low population of area A. Every effort must be made to compare figures on the same basis.

Another case may illustrate the misleading use of ratios and percentages. Sometimes we are asked to derive statistics (secondary statistics) from given tables of figures. For example, we may wish to calculate the percentage increases in unemployment over a number of years:

Table 8.3

Year	1	2	3	4
Unemployment	200 000	252 000	310 000	376 000

There are at least two ways in which we can calculate the percentage increase here:

Method 1.

Year 2 (increase on Year 1) = 26 per cent
Year 3 (increase on Year 2) = 23 per cent
Year 4 (increase on Year 3) = 21 per cent

Method 2.

Year 2 (increase on Year 1) = 26 per cent
Year 3 (increase on Year 1) = 55 per cent
Year 4 (increase on Year 1) = 88 per cent

In *Method* 1. where the percentage increase seems to be falling, the percentage rise is worked out using the year immediately preceding as the base year. In *Method* 2., the percentage increase is worked out using Year 1 as the base year for each calculation, and the rise appears to be growing rapidly! Once again, figures must be compared on the same basis, or else the method of working must be explained to the reader.

Some students have difficulty in averaging percentages when asked to derive statistics. The simple rule of 'weighting' (see Chapter 9) must usually be observed. In many examples, percentages are derived from different base figures, and if we wish to calculate an overall percentage from the percentage figures already worked out, we cannot simply take a straight average. An example may make this clear:

Table 8.4 Regional coal output (kg)

	Year 1	Year 2	% increase or decrease
Colliery P	30 000	30 500	+1.7
Colliery Q	15 000	16 000	+6.7
Colliery R	16 000	20 000	+25.0
TOTALS	61 000	66 500	+9.0

In this example it would be wrong to assume that the total (or overall) percentage increase is simply:

$$\frac{1.7 + 6.7 + 25.0}{3} = 11.1 \text{ per cent}$$

The correct method is to weight each individual percentage by its corresponding base year figure:

$$\frac{(30\ 000 \times 1.7) + (15\ 000 \times 6.7) + (16\ 000 \times 25.0)}{61\ 000} = 9.0 \text{ per cent}$$

This gives the same result, of course, as the calculation of the percentage increase in the totals:

$$\frac{66\ 500 - 61\ 000}{61\ 000} \times 100 = 9.0 \text{ per cent}$$

Use and interpretation

Accuracy

Strict accuracy is not to be found in any branch of applied human knowledge. Even in science, from the study of microbiology to interplanetary rocket-launching, exact measurement does not really exist.

In the social sciences, especially, we are hampered at the outset by the problem that the things we wish to measure are people, or the results of people's actions. As was suggested in Chapter 1, people and human events are much less predictable than atoms and molecules. One cannot subject people, exports or unemployment to controlled laboratory conditions and measure them or test them for reactions. Statisticians in business or government fields must rely on interviewers, the postman, and on individual bodies and agencies for pieces of information collected inside or outside the country, in all sorts of conditions and at various times. In this situation, accuracy must be very much a compromise, although we should try to make it a good compromise.

The Government would never dream of publishing a statement such as 'the value of goods and services exported in a certain year was £9 740 363 217.35'. And if it did, we could be certain of one thing—the figure would be wrong! Such accuracy, especially in the matter of exports, is unbelievably precise. Not only can we disbelieve the pence, but also the figures after the first four. It is extremely doubtful whether even the first four figures are within £100 million of the true figure! Yet, this seemingly large degree of possible error is quite acceptable if we compare it with figures of thousands of millions of pounds which make up our national accounts, of which exports is a part. In this situation we are well content with a useful figure which is taken to the 'nearest million pounds'. To define the figure more precisely would cost more than such a figure would be worth, if, indeed, it were possible.

Similarly, the Government found it necessary only to express the population of the UK as 55 848 thousand in 1981. This shows a greater degree of accuracy than in the previous case, and is presumably worth the far greater census expenditure entailed.

To summarize, we can say that the three limitations on accuracy are:

1. the *possibility* of greater accuracy.
2. the *desirability* of greater accuracy.
3. the *cost* of greater accuracy, in time and money.

Time, in the last respect, is often more valuable than money. Statistics collected probably at great cost may be useless if they are out of date by the time they are presented.

Many examination candidates, often in their desire to please, go beyond the requirements of the question in the matter of accuracy. This, for example, might take the form of working out more decimal places than are asked for, and is the way to lose marks. When no particular degree of accuracy is asked for, however, candidates must simply use their discretion, and work to a degree at least as accurate as the figures which make up the particular question.

Two reasons may be given for rounding numbers: **Rounding**

1. Most figures (see above) are approximations from the outset. Therefore, no great harm is done by rounding them, so long as we round all the figures in the same section to the same degree of accuracy.

Right	*Wrong*
361 000	361 000
428 000	427 885
39 000	39 000
2 000	2 000
TOTAL 830 000	TOTAL 829 885

2. The average reader (to whom we are likely to present our statistics) will be simply bemused, if not irritated, by figures running into millions which pretend to great accuracy.

£3 436m is much more readily appreciated than £3 436 428 376.

Errors

By errors in the sense used in the previous sections we do *not* mean mistakes in calculation or measurement. As has been said above, there are limitations to the degree of accuracy which we can achieve. Therefore, a compromise must be sought, and approximate figures agreed upon. Although the exact figures in an enquiry may not be known, almost certainly the margin, or range, within which the correct figure lies can be deduced. It is in this sense that a margin of error can be assigned to figures.

Percentages

As we suggested under Method, most of the percentages which the student will be asked to calculate will probably be concerned with report writing. The student is usually given a table of figures which can be compared much more easily by reducing some or all of them to percentages. As enough has already been said on percentage calculations some suggestions are offered on *report writing* in examinations.

Report writing

It is often said that there are three main rules to be observed in the writing of a report:

1. Accuracy.
2. Brevity.
3. Clarity.

Apart from the labour of deriving additional statistics which suggest the growth and performance of the original figures, the student must break down the report into sections.

Suitable sections might be:

1. A description of the changes and direction of change in the figures. As this is not the important interpretative part of the report, it should be extremely brief and, if the student wishes, may be left out completely.
2. Comments on the relationship and dependence of the figures on one another. If necessary, derived statistics, tables and charts could be used sparingly to illustrate such comments.
3. An explanation of why the figures have changed. Here the student's knowledge of economic or business affairs will be tested, as will knowledge of the events of the period under discussion.
4. An intelligent projection into the future, if this is possible and suitable. This section should not be too long because it may be regarded as part of the previous sections.

It is often a mistake to try to sum up at the end of such a report. A conclusion of this nature may distort your previous arguments in an attempt to condense them, and at best your conclusion may be mere repetition. Let each section stand on its own feet.

Titles and headings, references to the tables of figures, and your derived statistics, must be clear and full. A great deal depends on the student's narrative powers and on an understanding of statistics, and their everyday use. This is a searching type of question which demands more than sheer memory or the mechanical recitation of formulae.

Some tips for MINITAB users

When we are inside MINITAB and see the MTB > prompt we can add any remarks that we wish to our analysis by prefacing these comments with the hash ('#') symbol. The MINITAB package understands that what follows # is text rather than command. These comments will appear on screen and paper outputs. Comments can be useful in labelling output and/or reminding us of our logic and procedures, e.g.: **Making remarks**

```
MTB > # Plot of AGE and INCOME excludes 3 extreme values;
MTB > # Three extreme values of AGE suspected as errors.
MTB > PLOT C11 C12
```

For both micro and mainframe users of MINITAB there is a 'help' facility available whenever the MTB > prompt appears on the screen. To learn about the help facility, type HELP HELP and press ⟨**RETURN**⟩ or ⟨**Enter**⟩. For general information about the package, key HELP OVERVIEW. Most MINITAB commands have a help facility that describes their use. To obtain a list of the MINITAB commands available type HELP COMMANDS. To obtain aid with a particular command, e.g. PLOT, you would need to key HELP PLOT. To discover how to use a subcommand within a command, e.g. the XLABEL subcommand of PLOT, you will need to type HELP, then the command name, then the subcommand name: **MINITAB HELP**

```
MTB > HELP PLOT XLABEL
```

Try all these suggestions listed above! The help facility usually gives example uses, plus general grammatical statements. In MINITAB reference materials, 'C' refers to any column, 'K' to any constant, and 'E' to any column or constant. Square brackets, [ ], indicate optional elements or 'arguments' of keyed commands.

It is important to be clear about the various sorts of computer file with which we work. **Revision: MINITAB files**

Data can be keyed directly into the MINITAB worksheet using SET or READ commands, or (in the micro version) in the 'data editor' mode accessed by pressing ⟨**Esc**⟩.

103

Simple *data files* of the worksheet columns can be saved from within MINITAB with the WRITE 'file name' command, and they are later recalled to the worksheet with the READ 'file name' command. These simple data files contain nothing but column data (i.e. no variable name or stored constants), and are easily transferred to other software packages (e.g. spreadsheets). Both READ and WRITE commands require us to specify which columns in the worksheet are being used.

Complex combinations of data plus stored variable names and constants can be saved in a MINITAB *workfile* with SAVE 'file name'. A workfile is recalled to the worksheet with the RETRIEVE 'file name' command. Neither command lists the columns used.

The purpose of an *output file* is to collect and store, usually for later printing, not data but a complete record of our interaction with the package (i.e. commands plus responses). We open an output file with OUTFILE plus 'file name', and close it at the end of the session with NOOUTFILE.

We choose the file names ourselves for each of the three types of file listed above. In MINITAB, file names are always typed within single quotation marks. We should choose file names that make sense in the data analysis context (i.e. not comic file names). It often helps to number files, especially output files, e.g. 'GRAPH1', 'GRAPH2'. MINITAB automatically adds a suffix to the file name. This immediately identifies the file as a data file ('.DAT' or '.DATA') or a workfile ('.MTW' or '.MTWK') or an output file ('.LIS' or '.LIST'). Outside MINITAB, for instance when you want to print or rename one of these files, the operating system of your computer will insist that the full file name, including suffix, is given. Almost certainly, the operating system will expect to see the file name *without* single quotation marks.

Averages

It is important to be able to describe a group of items so as to distinguish it from some other group which might have many similar characteristics.

If we were asked to contact Mrs Jones, in a crowd of people, it would obviously help if we knew that she was the very tall woman, the very thin one, or possibly the one whose left shoulder drooped. The more details given about her, the easier she would be to identify. In this chapter, we shall concern ourselves with the 'height' of Mrs Jones, leaving her other features to Chapter 10.

The average value of a set of grouped or ungrouped data may be looked at in two ways. It is a method of picking out a typical or representative item from the group. It is also a measure of central tendency, i.e. a point round which the data tend to locate themselves. On a graph, it serves to fix the position of the curve, in relation to the horizontal scale.

There are several types of average which can be used to summarize a set of data. These are discussed below.

The arithmetic mean (usually referred to as 'A.M.') is the best-known type of average, and it is popularly used to summarize a batsman's or a footballer's score. It replaces a long list of runs or goals scored throughout the season. **The arithmetic mean**

UNGROUPED DATA

The *simple arithmetic mean* for ungrouped data may be expressed by the fraction:

$$\text{A.M.} = \frac{\text{total value of items}}{\text{total number of items}} \qquad \text{(F.9.1)}$$

Thus, if during a season a batsman scores 22, 8, 0, 14, 5, and 17 runs in six innings—applying the formula, we get:

$$\text{A.M.} = \frac{22 + 8 + 0 + 14 + 5 + 17}{6} = \frac{66}{6} = 11 \text{ runs}$$

Note 1: The zero score must be included as an item in the denominator (six items altogether).

Note 2: Although during six innings the batsman never *actually* scored 11 runs, nevertheless, the scores above and below 11 are exactly balanced round this figure.

GROUPED DATA

Where we have grouped data to deal with, i.e. where the items are numerous and are arranged in class intervals, we must use a *weighted arithmetic mean* rather than a simple arithmetic mean.

The short table below shows the scores of 15 batsmen:

Table 9.1

Runs scored	Number of batsmen
6	4
8	4
12	5
17	1
34	1
TOTAL	15

The class interval here is in single runs. It is obviously incorrect simply to add the 'Runs scored' column, and divide by the total 'Number of batsmen', or even to add the 'Runs scored' column and divide by five (the number of class intervals). Students frequently make such careless mistakes. Each class interval must first be multiplied by its corresponding frequency (Number of batsmen) in order that due weight be given to each score.

Table 9.2

1 Runs scored	2 Number of batsmen	3 Column 1 × Column 2
6	4	24
8	4	32
12	5	60
17	1	17
34	1	34
TOTALS	15	167

The 'Total value of items' (167 runs) is then divided by the 'Total number of items' (15) to give:

$$\text{A.M.} = \frac{167}{15} = 11.13 \text{ runs}$$

This is the correct method to use when the class interval consists of a *single figure in a discrete series*.

When we deal with grouped data where the class interval consists of values over a range (the more normal case), the method is slightly different.

Consider the table given in Chapter 7:

Table 9.3 Weekly bonus earnings in an industrial region

Weekly bonus earnings (£s)	Number of workers (thousands)
20–29.99	52
30–39.99	256
40–49.99	170
50–59.99	68
60–69.99	30
70 and over	24
TOTAL	600

The midpoints of the class intervals are found, and these midpoints are taken to be rough averages, representative of each class. We are making the assumption here that the number of items in each class is spread evenly round the midpoint; i.e. the average value in the first class interval is £25. This is the best we can do lacking more detailed information, in spite of the fact that our assumption may be wrong. In fact, our assumption *is* wrong! If the student will look at Table 7.2 it can be seen that, in this particular example, the first class interval was weighted heavily with items at its upper end. Nevertheless, with only the simple table above to go on, this is the assumption we must make.

Each midpoint is multiplied or weighted by the corresponding frequency, as follows:

Table 9.4 Weekly bonus earnings in an industrial region

1 Weekly bonus earnings (£s)	2 Number of workers (thousands)	3 Midpoints	4 Column 2 × Column 3
20–29.99	52	25	1300
30–39.99	256	35	8960
40–49.99	170	45	7650
50–59.99	68	55	3740
60–69.99	30	65	1950
70 and over	24	75	1800
TOTALS	600	—	25 400

The total of column 4 gives the 'Total value of items' in our previous formula for the arithmetic mean. All that remains is to divide this figure by the 'Total number of items', i.e. the total of column 2.

Note. The figures for 'Number of workers' are in thousands, and therefore, strictly speaking, the figures in column 4 should have been multiplied by 1000, to give 1 300 000, 8 960 000, etc. In practice we can ignore this point, because we would merely be multiplying and dividing by 1000. Thus,

$$£\frac{25\ 400\ 000}{600\ 000} = £42.33$$

It should be realized that this answer is, at best, only an approximation. There are several reasons for this:

1. 'Number of workers' figures are presumably given to the nearest 1000.
2. Actual earnings figures may not be spread evenly over the class intervals—it would be quite remarkable if they were! Therefore, the midpoint figures may not be suitable.
3. The last open-ended class is presumed to end at £80—this, in fact, may not be true.
4. It is hoped that any errors made in compiling the table may be compensating—particularly in the case of the figure '600 000 workers'. It is probable that, as only six classes have been taken, there might be considerable error in this figure.

Short methods of calculating the A.M.

Where the student meets with lengthy or complex frequency tables, a short-cut may be adopted by using the 'arbitrary origin' or, as it is commonly called, the 'assumed mean' or 'working mean'.

UNGROUPED DATA

In our first example in this chapter, we were given the figures for a batsman's scores as follows:

$$22, 8, 0, 14, 5, 17$$

With such a small number of items, the arithmetic mean is easily found by the ordinary method, but we will repeat the example to show the basic principle.

We first assume an arithmetic mean, say, 10 runs. We then take the deviations of each actual figure from the assumed mean, and give our results an appropriate plus or minus sign (Table 9.5).

The balance of the plus and minus deviations is +6. We find the average of this total deviation by dividing it by the total number of innings (six):

$$\frac{+6}{6} = +1$$

Table 9.5

Runs scored	Deviations from assumed A.M. of 10 runs	
	+	−
22	12	
8		2
0		10
14	4	
5		5
17	7	
TOTALS	+23	−17

Finally, we add this result (or subtract if it is negative) to the assumed mean:

$$10 + 1 = 11 \text{ runs}$$

Thus, we find the same answer as before. The student must not think that, because we have 'guessed' the assumed mean, there is any guesswork or inexactness about this answer. The answer is absolutely accurate, and the 'estimated' mean of 10 was revised to the correct mean of 11 in the course of the calculation.

GROUPED DATA

We can apply the same principle to our first example in grouped data (Table 9.1), in which the arithmetic mean score of the batsmen was calculated as 11.13 runs. Let us assume a mean of 10 runs. Our table is then set down as follows:

Table 9.6

1	2	3		4	
Runs scored	Number of batsmen	Deviation from assumed mean of 10 runs		Column 2 × Column 3	
		+	−	+	−
6	4		4		16
8	4		2		8
12	5	2		10	
17	1	7		7	
34	1	24		24	
TOTALS	15	—	—	+41	−24

The balance of the plus and minus weighted deviations is therefore +17. Taking the average of this deviation for the 15 batsmen, we get:

$$\frac{+17}{15} = +1.13$$

This result is added to the assumed mean:

$$10 + 1.13 = 11.13$$

which is the same answer we obtained before.

A similar short method is given below for the example in Table 9.3. The midpoints have been calculated as before, but instead of setting down 25, 35, 45, etc., we have decided to take £45 as the assumed mean (this being a convenient midpoint), and to record this in the midpoint column as 0. Because the remaining midpoints above and below £45 are all at intervals of £10, we can work in units of £10, which will reduce the deviations to single digits. For example, the midpoint of £65 is £20 more, which is two units of £10, or +2. Deviations below the £45 are given a minus sign.

Table 9.7 Weekly bonus earnings in an industrial region

1 Weekly bonus earnings (£s)	2 Number of workers (thousands)	3 Midpoints deviation from £45 (units of £10)	4 Column 2 × Column 3 deviations
20–29.99	52	−2	−104
30–39.99	256	−1	−256
40–49.99	170	0	0
50–59.99	68	+1	+68
60–69.99	30	+2	+60
70 and over	24	+3	+72
TOTALS	600	—	−160

$$\text{The true arithmetic mean} = £45 + £\left(-\frac{160}{600} \times 10\right) = £45 - £2.67 = £42.33$$

It can be appreciated that this simplification reduces paper arithmetic to mental arithmetic, as well as cutting out one calculation (third row down) altogether. But it is essential to be careful about the correction, i.e. to balance out the plus and minus deviations, obtain the average of the result, multiply this by the class interval (£10 in our example), and either add to, or subtract from, the assumed mean.

Note. The correction $£\left(-\dfrac{160}{600} \times 10)\right)$ which we make to the assumed mean must be fully worked out before we subtract it from, or add it to, the assumed mean. For this reason the correction is put in brackets.

A point of interest in calculations using the assumed mean is that if the plus and minus deviations were to cancel out each other *exactly*, this would mean that we had chosen, by chance, the true mean as our assumed mean. This illustrates the idea of the arithmetic mean, which is a point in a distribution where plus and minus deviations of the individual items cancel each other out.

Another well-known average, the mode, is best remembered by the French phrase *à la mode*, meaning 'in the fashion'. In a distribution, the model average is the most fashionable value among the items. 'Fashionable' is taken to mean 'most frequently occurring', so that the modal average is the value which most frequently occurs among all the items in the distribution. For example, in Table 9.6, listing runs scored by 15 batsmen, the mode is 12 runs, because five batsmen made this score, while the other scores were achieved by fewer batsmen in each case.

The mode

It is usual to refer to the 'modal class' where data are presented in class intervals containing a range of values. Thus, in the last example (Table 9.7), the modal class is the second, 30–39.99, because 256 000 of the 600 000 items fall into this class.

On a frequency curve, the mode is easily found, because it is the highest point of the curve. This is considered in more detail in the next chapter. Unless the values of all the items in a distribution are known, it is not possible to determine the mode with precision. However, two methods may be used to find an approximate value for the mode.

GRAPHICAL METHOD

The modal value may be measured by drawing a vertical line through the highest point of the frequency curve and reading off the value on the *X*-axis. If the information is limited to a frequency distribution, then the modal value may be found by the use of a histogram. We find the modal class, i.e. the highest column of the histogram, and join the top corners of this column diagonally to the adjacent corners of the columns on either side. This is shown below. This latter method is valid only when the class intervals of the modal class and the two classes (one on each side) are equal.

ARITHMETIC METHOD

If we take the example of the weekly bonus earnings (in Table 9.7), we have already seen that the modal class is the second down, 30–39.99. Obviously, the mode is somewhere within this class. A fair guess at the mode might be the midpoint of this class, i.e. £35. If we examine the distribution of all the items, however, it will be noticed that the frequencies rise sharply at the lower values and tail off more gradually at the higher values. This would suggest that the

111

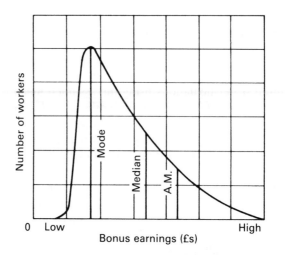

Fig. 9.1 Weekly bonus earnings in an industrial region

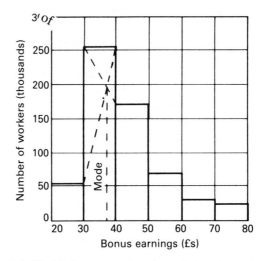

Fig. 9.2 Weekly bonus earnings in an industrial region

mode is *more* than £35, i.e. between £35 and £40. This conclusion is drawn because the class interval 40–49.99 contains more items (170) than the class interval below the modal class, which contains only 52 items. The distribution is more heavily weighted or pulled to one side, in favour of the higher values within the modal class 30–39.99. By considering the classes on either side of the modal side, it is easy to calculate the extent of this pull.

The three classes in question are extracted from the table together with their frequencies:

Table 9.8

Weekly bonus earnings (£s)	Number of workers (thousands)
20–29.99	52
30–39.99	256
40–49.99	170

First, we know that the mode is between £30 and £40, and nearer to £40. If we now subtract the two frequencies on either side of the modal frequency from the modal frequency itself, we shall get the results:

$$256 - 52 = 204$$
$$256 - 170 = 86$$

The extent of the pull above £30 can now be measured as follows:

$$\text{Mode} = £30 + \frac{204}{204 + 86} \times £10 = £30 + £7.03 = £37.03$$

To the lower end of the modal class (£30) we add the proportion of the class interval (£10) which has been divided in the ratio of 204:86—the differences in frequencies between the classes on either side of the modal class and the modal class itself.

UNGROUPED DATA OR GROUPED IN SINGLE UNITS

The median

The *median* is a third kind of average which is widely used. To find this average it is necessary to arrange all the items in a distribution in either ascending or descending order, and to pick out the middle item of the array. This is the median item, and its value is the median value. This is shown in the table below:

Table 9.9 Factory size by number of employees

Number of employees	Array of number of employees
516	22
22	48
343	71
197	91
71	197
1245	343 ← Median
507	507
48	516
661	661
91	672
672	1245

In this array of 11 items, item number 6 is the middle item because it has an equal number of items above and below it. Of course, it is perfectly possible to pick out a middle item when the total number of items is odd. When we have an even number of items there will obviously be two middle items with an equal number of items above and below them. To find the median value here we must take the arithmetic mean of these two middle items. In the example above, we can imagine another large item being added to bring the total to 12 items. The two middle items will then be 343 and 507. The arithmetic mean of these is:

$$\frac{343 + 507}{2} = \frac{850}{2} = 425$$

It can be seen that this median value does not correspond to any actual figure in the array. Thus, in ungrouped data such as that in our table, we have the rule that, if the number of items is odd, the median value will be an *actual value*, while if the number of items is even, the median value will be an *estimated value*. A useful formula for finding the position of the median when the data is either ungrouped or simply grouped in single units (as is usually the case with discrete units) is:

$$\text{Position of median} = \frac{N + 1}{2} \qquad \text{(F.9.2)}$$

After the position of the median has been found it is always necessary to find its value.

The table below shows a discrete series (number of rooms) in grouped form:

Table 9.10 Housing types in a district

Number of rooms	Number of houses in type	Cum 'less'
3	38	38
4	654	692
5	311	1003
6	42	1045
7	12	1057
8	2	1059
TOTAL	1059	

In this table, a cumulative 'less than' column has been added. Applying the formula F.9.2, we get:

$$\frac{1059 + 1}{2} = \frac{1060}{2} = 530$$

The 530th item in the cumulative 'less than' column obviously lies in the four-roomed house type. Therefore the median value is four rooms.

DATA GROUPED IN CLASS INTERVALS CONTAINING A RANGE

Where data has been classified in intervals each containing a range of values, the position of the median is found by applying the slightly different formula:

$$\text{Position of median} = \frac{N}{2} \qquad \text{(F.9.3)}$$

To illustrate this we take again the table of weekly bonus earnings:

Table 9.11 Weekly earnings in an industrial region

Weekly bonus earnings (£)	Number of workers (thousands)	Cum 'less'
20–29.99	52	52
30–39.99	256	308
40–49.99	170	478
50–59.99	68	546
60–69.99	30	576
70 and over	24	600
TOTAL	600	

Applying the formula $\frac{N}{2}$ (F.9.3), we find the position of the median to be $\frac{600}{2} = 300$, from either end of the distribution. From a glance at the cumulative column it can be seen that the median will lie at the *upper end* of the class interval '30–39.99'. Once again we must assume that items within each class are distributed evenly throughout the class. The median value is now found by interpolation, as follows:

Below the class interval '30–39.99' there are 52 items. Therefore, to find the value of the median, the 300th item, we must continue counting for a further 248 items from the bottom of class interval '30–39.99'. In fact, because there are 256 items in the latter class, we must proceed $\frac{248}{256}$ ths of the way up this class. This median value will be, therefore, $£30 + \frac{248}{256} \times £10 = £30 + £9.69 = £39.69$.

As in the case of the mode, the value of the median may be found graphically from the cumulative frequency curve (Fig. 9.3).

Two ogives have been drawn. In one, the cumulative totals have been calculated in *ascending* order of the groups, by value (less than), while the other is in *descending* order (more than). For example, there are 548 workers earning

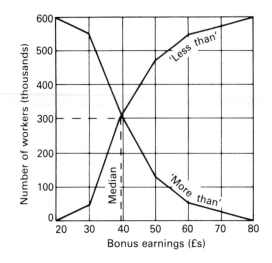

Fig. 9.3 Cumulative curves (ogives) of weekly bonus earnings in £s

more than £30, and 292 earning more than £40, and so on. It is really only necessary to draw one of the curves and then draw a horizontal line from the median position (300) and drop a perpendicular to the *X*-axis to find the median value. The two curves show that the position and value of the median can be found from the point where they intersect.

The geometric mean

A fourth kind of average, which is not as frequently met with, is the *geometric mean* (G.M.). To find the geometric mean of a number of items, the values are multiplied together and the root of their number taken. If a number of items had the values 75, 11, 17, 24, for example, the formula for the geometric mean would be $(75 \times 11 \times 17 \times 24)^{1/4}$.

This can more easily be calculated by the use of logarithms (Table 9.12) or on a calculator with a suitable facility.

Table 9.12

Value of item	Logarithm
75	1.8751
11	1.0414
17	1.2304
24	1.3802
TOTAL	5.5271

We divide this logarithm by four (i.e. the number of items), to get the fourth root, $\dfrac{5.5271}{4} = 1.3818$. The anti-logarithm of this figure gives the answer 24.08 as the geometric mean.

More advanced calculators usually enable the student to work out a fractional power (or *n*th root).

This is calculated by first converting each number or value into a reciprocal (i.e. one divided by the number).

We next find the simple arithmetic mean of these reciprocals, which will, in general, be fractions.

Having found this, we write down its reciprocal, which means that we reverse the numerator and denominator of the answer. This gives us the *harmonic mean* (H.M.).

The harmonic mean

Example:

Find the harmonic mean of the following numbers: 12; 8; 24; 6.

1. The reciprocals are $\dfrac{1}{12}, \dfrac{1}{8}, \dfrac{1}{24}$, and $\dfrac{1}{6}$.

2. Because there are four items, the simple arithmetic mean is their total divided by four:

$$\frac{1}{4}\left(\frac{1}{12} + \frac{1}{8} + \frac{1}{24} + \frac{1}{6}\right) = \frac{1}{4}\left(\frac{10}{24}\right) = \frac{10}{96}$$

3. The reciprocal of this is $\dfrac{96}{10} = 9.6$.

The corresponding arithmetic mean is 12.5.

Use and interpretation

The five kinds of average already described possess certain advantages and disadvantages in use. Some are more suitable than others, according to the type and kind of data to which the statistician wishes to apply them. Often the wrong kind of average is used either deliberately or carelessly.

Desirable qualities in an average

1. Easy to calculate.
2. Easy to understand.
3. Suitable for arithmetic treatment, i.e. it should be flexible in other uses.
4. It should preferably be based on the values of all the items in the distribution.
5. It should be an exact value, not an estimated one.
6. It should not be unduly distorted by the influence of extremely high or low values.

The use of this particular average is so general and widespread that it is pointless to try to list its various applications. There is, indeed, a dangerous assumption on the part of the public that the word 'average' *means* what is

The arithmetic mean 117

known to the statistician as the arithmetic mean. This, of course, is wrong. Every reference to an average should say what kind of average is meant. Unfortunately, because the person in the street 'cannot be bothered' with such technical details he or she often becomes the victim of trickery and deceit.

ADVANTAGES

1. The calculation of the arithmetic mean is widely understood.
2. The calculation is not complicated, although it may often be more lengthy than that of the other averages.
3. The value of every item in the distribution is included, and although a few extremely high or low values may distort the average as a measure typical of the distribution, nevertheless, there is an arithmetical exactness about the figure which is often missing in, say, the mode or the median.

DISADVANTAGES

1. A few items of very high or low values may make the average rather unrepresentative of the whole distribution.
2. The arithmetic mean cannot be measured or checked by graphical methods.
3. The figure is not likely to correspond to any actual value in the distribution itself.

The mode   Sometimes one finds statements such as 'The average family in this country has 2.317 children'. To quote the arithmetic mean in such a case makes the advantage of arithmetical exactness appear rather comic. If a journalist were writing for the layperson it would be better to forego the air of exactitude and give the mode. This figure would really be quite as exact and, at the same time, preserve the atmosphere of good, solid sense.

The modal average has many uses in business and government, e.g. the builder of housing estates obviously wishes to know the modal size of family in order to know the modal number of rooms. The arithmetic mean is worthless to him because, although the modal average family may exist, the arithmetic mean average family certainly does not. An army tailor or bootmaker works (often with distressing effect!) to modal sizes because, not only is this average an actual value, but often it covers the actual value for the majority of the distribution. Many people complain that too rigid acceptance of modal sizes in manufacturing and production produces lack of flexibility and poor fit in many products.

ADVANTAGES

1. The mode is not unbalanced (made unrepresentative) by extreme values.
2. The mode is an actual value, and often represents the majority of cases.
3. The values of all the items in the distribution need not be known in order to calculate the mode.
4. It is easy to understand, and is usually readily appreciated on a graph.

1. Where a distribution is widely dispersed (see Chapter 10) over the range, and especially if the distribution has two or more peaks (*bimodal*), the mode becomes less useful as an average.
2. Although the mode is particularly useful as the average of a discrete series, its value may be only approximate in a continuous series.
3. Because of its lack of exactness (particularly in a continuous series), the mode is unsuitable for other kinds of calculation, as is the arithmetic mean.

The median

The median is not as widely used as the two previously described averages. We use it most frequently when we require a measure of location which is not affected by high or low value items, and when we wish to measure the change in different sets of distributions which move in a similar direction in a similar manner. Thus, the median, which divides a distribution in half by number of items, is frequently used as an average in testing general abilities. Examples are intelligence, educational scores and tests. Other uses can be found in the measurement of changes in the cost of living, when the prices of many items of general expenditure in all parts of a country are collected and reduced to a single figure.

The journalist's joke that 'Fully half the children in the country are below average intelligence', may be sensational, but it is also exact if the 'average' is the median. In Chapter 10 (Dispersion), the median receives further mention because this average is particularly useful when we wish to divide up the range of values in a distribution into fractional parts, to find out the characteristics of each part.

ADVANTAGES

1. Extremely high or low values do not distort it as a representative average.
2. It is readily obtained, even if we do not know the values of all the items. Also it is unaffected by irregular class intervals or open-ended classes, i.e. we do not have to estimate for these.
3. Unless we are dealing with class intervals which each contain a range of values, the median is an actual value.

DISADVANTAGES

1. The median gives the value of only one (the middle) item, although the surrounding items may have the same value. But if the number of items is few, or if the items are spread erratically above or below it, the median may lose its value as a representative figure.
2. In a continuous series, grouped in class intervals, the value of the median can only be estimated.
3. The median is not suitable for arithmetic treatment in advanced work.

The geometric mean

The main uses of the geometric mean are to be found in cases where we wish to measure changes in the rate of growth, e.g. where the magnitude of one quantity depends directly on a previous magnitude. The population size at any point in time, for example, depends on the population which preceded it, because not only do births depend on the number of married couples in the previous period, but deaths will be largely dependent on the number of old people existing in the previous period.

Example:

If the population in a country in 1900 was 10 million persons, and the population in 1960 was 15 million persons, estimate the probable population in 1930.

The arithmetic mean estimate would be $\dfrac{10 + 15}{2} = 12.5$ millions.

The geometric mean estimate would be $(10m \times 15m)^{1/2}$ which, by logarithms is:

Number	Logarithm
10m	7.0000
15m	7.1761
TOTAL	14.1761

$$\frac{14.1761}{2} = \text{antilog } 7.0881 = 12.25 \text{ millions}$$

This is lower than the arithmetic mean estimate, and is more likely to be correct. The reason for this is that a population tends to grow in *geometric proportion*—an ever-increasing amount—and the use of an arithmetic mean here would assume a constant amount. This is best illustrated by means of a graph showing the arithmetic and geometric curves (Fig. 9.4).

In Fig. 9.4, the arithmetic (constant amount) curve is *above* the geometric (increasing amount) curve at all points between 10 m and 15 m. The estimated geometric mean will always be *lower* than the estimated arithmetic mean in such a case. See Chapter 7 for differences between arithmetic and geometric curves, and Chapter 19 for a comparison of arithmetic and geometric series.

ADVANTAGES

1. Because the geometric mean shows the *rate of change* on a basic quantity, this average is often used to calculate index numbers (see Chapter 13), e.g. where changes in prices are calculated and averaged on a points or percentage basis of the previous year.
2. Every item in the distribution is included in the calculation.
3. Extremely high values have not the disproportionate effect on the geometric mean as they have on the arithmetic mean (but *low* values have *more* effect on the geometric mean than on the arithmetic mean).

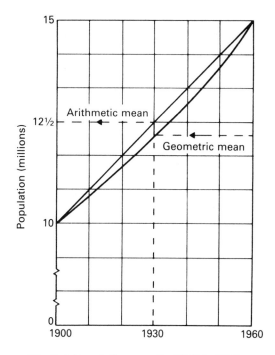

Fig. 9.4 Population growth, 1900 to 1960

DISADVANTAGES

1. The geometric mean is not easily understood (why should we multiply the items and take a root of the product?), and its function as an average is not as readily apparent as in the cases of the other averages.
2. It is not possible to calculate the geometric mean if the value of any item in the distribution is either zero or negative.
3. The value of the geometric mean may not correspond to any actual value in the distribution.

The harmonic mean

This average is of minor importance in statistics, being restricted to the averaging of *rates* as opposed to simple values, e.g., in kilometres per hour, revolutions per second, etc.

Thus, if a car went from A to B at 30 km/h, and back again at 20 km/h, the average speed for the whole journey would be given by the harmonic mean of the speeds. This is 24 km/h. Most people would assume the answer to be 25 km/h, and this is because they have used the wrong kind of average, i.e. the arithmetic mean.

Students often have difficulty in deciding which of the two averages to use. If the problem does not involve rates, speeds, etc., then the harmonic mean does not come into it. Even if rates *are* involved, it does not follow that the harmonic mean is the correct average to use.

121

Taking kilometres per hour as our example, one can say that if the kilometres are constant, then the correct average is the harmonic mean (as above). If the time (in hours) is constant, then the correct average is the arithmetic mean.

Thus, if the car in the above example had travelled, say, for one hour at 30 km/h and one hour at 20 km/h, then the average speed for the whole period of time would be 25 km/h (the simple arithmetic mean).

Misuse of averages

Suppose, in a certain country, there is industrial conflict between workers and employers. The trade union is seeking a wage rise and the employers' federation will not agree to this demand. Both sides are seeking to impress the public with the justice of their cases.

The union emphasizes the degrading poverty of its members in relation to other workers and in the light of the rising cost of living, and it asserts 'The average weekly bonus paid this year in the industry was under £15.'

The federation, anxious to make public the brazen manner in which they and the public are being held to ransom, states that 'The workers in the industry received last week an average weekly bonus of £51'.

Which side is the layperson to believe in the face of such staggeringly different figures? Neither? Is it a question of the reader's politics or newspaper, or whether the reader is a worker or an employer?

If a statistician, the reader will realize that *both figures are likely to be right*. But certain questions will occur to the statistician.

QUESTIONS TO BE ANSWERED

We will assume that each side has made a completely accurate calculation.

1. Each statement mentions an average, but does not *name* the kind of average. Can we imagine the kind of distribution of earnings in the industry, i.e. the size of the frequency at each level of earnings? It would probably follow a curve similar to the frequency curve illustrated in Fig. 9.5.

   This is much the same kind of curve as Fig. 9.1, and for much the same reasons, e.g. most workers—including part-time workers, and the majority of unskilled and semi-skilled workers—are paid earnings in the lower brackets, and the relatively fewer skilled workers and supervisors receive the higher incomes. Note that the axes are scaled but we presume that the reader would not see this graph nor have information on its details.

   On the curve are shown the typical positions in which the mode, median and arithmetic mean would appear (see Fig. 9.1).

   It is possible, therefore, for the union to understate the workers' reward by selecting the modal wage as the 'average weekly earnings', this being the lowest average which offers itself. Actually, the vaguenes of 'under £15' seems to suggest a modal class interval.

   The employers, on the other hand, might choose the arithmetic mean in order to prove high 'average earnings'. The precision of '£51' suggests an arithmetic mean (an added advantage is that it has an air of accuracy which suggests careful and final deliberation).

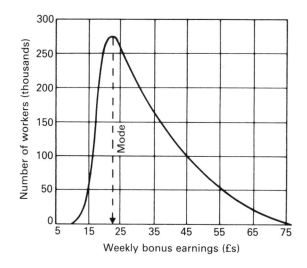

Fig. 9.5

Both the figures, therefore, are averages, but the public is not aware of the technical differences which make them non-comparable. To choose an average like this to suit your particular case might result in a difference, but probably not so large a difference as between the two quoted figures. We must look for other reasons.

2. The union might have included part-time workers' earnings in its calculations in order to weight its average on the low side. The employers might have deliberately excluded such classes and taken full-time workers only.

3. The employers speak of the earnings received 'last week'. This may have been a week of full work with good overtime on piecework, etc. The union speaks of the wage paid 'this year'. This may include the slack winter period when short-time piecework was the rule. Such differing bases would tend to inflate and deflate the averages respectively.

4. Finally, there may be differences in the definitions of 'earnings', 'industry', etc. (see Chapter 21 for a discussion of this).

Unless the union and the federation have calculated correctly, and in the same manner, the figures they arrive at will be different, and the results will be more or less worthless for purposes of comparison. Their efforts will only have been 'successful' if the public has been deluded in one way or the other.

Computing spreadsheets in statistics

Users of statistical packages should refer to subsequent chapters for the commands that calculate averages. These are Chapter 10 (MINITAB, SPSS^x) and Chapter 11 (SPSS/PC+).

Statistical packages

Convenient spread-sheets

Here we begin work with spreadsheets, reworking the cricket example in Table 9.1. The most popular spreadsheets look very similar to one another. Here, we display **Lotus 1-2-3**, but the spreadsheet **AS-EASY-AS** is virtually identical, and very few changes are required to make the example successful in the spreadsheet **Works**. Variations in commands for Works are also given here.

Chapter 3 contains a description of the general style of spreadsheets. You should reread the appropriate section before continuing. In the same chapter you should also refresh your memory about using *menus* and *highlight bars*. The most important point to remember is that in the spreadsheet *worksheet* we can type any combination of data, labels and formulae. We get instant answers to the formulae, with immediately revised answers, if we change the data. If you have a spreadsheet, get to know it by copying our examples. But first, read to the end of this chapter.

If you are using a PC and have Lotus 1-2-3, the package is most likely accessed by changing to the 'Lotus' directory, and then typing 123. Typical commands to call AS-EASY-AS are given on the right. Commands to access the two packages may vary slightly between installations.

```
cd lotus (RETURN)          cd aseasy4 (RETURN)
123 (RETURN)               aseasy (RETURN)
```

We should immediately enter the empty worksheet.

Works is an integrated package of spreadsheet, wordprocessor and database. On a PC the most likely command sequence is:

```
cd works (RETURN)
works (RETURN)
```

From the Works menu that appears, first select the 'Create New File' option with (**RETURN**) or (**N**), and then highlight the 'New Spreadsheet' option (with the cursor 'arrow down' key) from the next menu. Press (**RETURN**). Alternatively, menu options may be selected in Works (without moving the highlight bar) via the key letter indicated—here (**S**) for 'spreadsheet'. We then enter the worksheet and begin typing.

DATA, FUNCTIONS, FORMULAE

For all three spreadsheets we simply type the labels ('RUNS', 'BATSMEN', 'R × B') and cricket data from Table 9.1, plus the needed formulae, into the cells A1 to D13 (inclusive) as shown below in Fig. 9.6. Our inputs are into columns A, B and D only. We have left column C clear. The cursor highlights the 'active' cell. What we type into the active cell appears first in the command space at the top left of the screen, after the cell's column and row designation, e.g. D13. What we have keyed only appears in the cell itself after we have pressed one of the cursor arrow keys or (**RETURN**). For the moment ignore the new data in column G.

D13:	+D9/B9

	A	B	C	D	E	F	G
1	Runs	Batsmen		R × B			Batsman A
2							
3	6	4		+A3★B3			23
4	8	4		+A4★B4			15
5	12	5		+A5★B5			0
6	17	1		+A6★B6			27
7	34	1		+A7★B7			4
8							8
9	TOTAL1	@SUM(B3..B7)		@SUM(D3..D7)			12
10							3
11	TOTAL2	+B3+B4+B5+B6+B7					45
12							
13	MEAN			+D9/B9		MEAN GAVG(G3..G11)	
14							
15							

Fig. 9.6 Runs and batsmen example as keyed into the spreadsheet

We have chosen to calculate the total number of batsmen in two different ways. Each way involves typing a formula into a cell. The most obvious way to calculate the total is to use the built-in SUM function plus the *range* of cells over which we wish the sum to be made. To get the sum of the batsmen data in cells B3 to B7, we key into cell B9 the formula '@SUM(B3..B7)' (*without* quotation marks). In Lotus 1-2-3 and in AS-EASY-AS, all built-in functions begin with the '@' symbol. The Works equivalent is '=SUM(B3:B7)', again without quotation marks. Note that in Works the range is specified with a colon rather than full stops. In Works, functions and formulae begin with the 'equals' symbol ('=').

Cell B11 illustrates how we can arrive at the same total by creating our own rather longer formula. In Lotus 1-2-3 and AS-EASY-AS, formulae beginning with cell designations start with the positive (+) sign (or negative (−) sign if appropriate to the formula). We therefore key into cell B11:

$$+B3+B4+B5+B6+B7$$

This does the same job as the @SUM or =SUM function. The formula immediately above is also acceptable in Works, or alternatively we can type =B3+B4+B5+B6+B7 in cell B9 of the Works worksheet.

Continue to fill the cells with the cricket data and the functions and formulae. In these three spreadsheets the arithmetic operators are '+', '−', '/' (division), and '✲' (multiplication). Exponentiation is indicated by '^'. The formula 5^2 (5 squared) will give the result 25. In our example the average number of runs is calculated in cell D13 by dividing the total runs scored (held in D9) by the total number of batsmen (held in B9 or B11)—i.e. +D9/B9.

As you type, remember that what is keyed will appear on the screen in the top left corner. The active cell in the worksheet below is only actually filled when we move to another cell with ⟨**RETURN**⟩ or the cursor arrow keys. Also, what actually fills the cells are not the functions and formulae but the *results* of the calculation. What actually appears in the worksheet as a result of keying Fig. 9.6 is shown below:

D13:	+D9/B9

	A	B	C	D	E	F	G
1	Runs	Batsmen		R × B			Batsman A
2							
3	6	4		24			23
4	8	4		32			15
5	12	5		60			0
6	17	1		17			27
7	34	1		34			4
8							8
9	TOTAL1	15		167			12
10							3
11	TOTAL2	15					45
12							
13	MEAN			11.13333		MEAN	15.22222
14							
15							

Fig. 9.7 Runs and batsmen example as displayed on the screen

We have introduced more fictitious data for 'Batsman A' into column G in order to demonstrate that these three spreadsheets have other built-in functions. Here we use @AVG (or =AVG in Works) to calculate the arithmetic mean of a range of data. Other statistical functions include @COUNT (=COUNT) and @STD (=STD). This last function calculates *standard deviation* (see Chapter 10).

SPREADSHEET FACILITIES

All three spreadsheets have a 'help' facility that is called up with ⟨**F1**⟩ and switched off with ⟨**Esc**⟩. In Works, pressing ⟨**F1**⟩ also gives access to an excellent 'Works Tutorial' program. Other spreadsheet commands are accessed by pressing ⟨**Alt**⟩ (in Works) or ⟨/⟩ (in Lotus 1-2-3 or AS-EASY-AS). These commands appear as a menu bar running across the top of the screen, or down the side (AS-EASY-AS). As each command is highlighted (using the cursor keys) a short description of the command's contents appears at the top of the screen. The highlighted command is actually selected with ⟨**RETURN**⟩. This usually leads us into further sub-menus. Selecting the appropriate menu option we can do such tasks as store spreadsheets ('File') or leave the package entirely ('Quit', 'Exit'). In Works, the 'Exit' option is accessed by first selecting 'File'. If we wish to stay in the spreadsheet package we turn off the menu system with

⟨**Esc**⟩. If you get lost in the menu system try pressing ⟨**Esc**⟩ several times to get back to the normal worksheet operation.

In all three packages, pressing ⟨**F2**⟩ takes us to the 'edit' mode in which mistakes in the active cell can be corrected with the use of cursor and ⟨**Del**⟩ keys rather than complete overtyping. The ⟨**F2**⟩ edit mode is switched off with ⟨**Esc**⟩. In all three spreadsheets this edit mode can also be accessed in the command menu called up with ⟨**Alt**⟩ (Works) or ⟨/⟩ (Lotus 1-2-3, AS-EASY-AS).

If you are using a spreadsheet for your statistical work, and have no specialist statistical package available, you will find that you can calculate all your needed statistics by keying formulae into the cells. Whereas statistical package users may be able to calculate statistics like *correlation* (Chapter 11) with a single command, spreadsheet users may well have to translate the book formulae into the sort of spreadsheet terms outlined in the cricket example above. Some spreadsheets allow easy calculation of *regression* (Chapters 11 and 12), e.g. with the 'Data' option. Graphs and charts can be produced with 'Graph' (see Chapter 18). ('Data' and 'Graph' are accessed in the command menu with ⟨**Alt**⟩ or ⟨/⟩). Use the ⟨**F1**⟩ help facility to find out what is available, and how it can be called.

10

Dispersion

Method

In Chapter 9 we discussed one method of describing our imaginary Mrs Jones—namely, by mentioning her height. If, however, there are other individuals present of similar height, this will not serve to identify her, and we must mention some other feature which will distinguish her, such as the fact that she is thin, or that she is lop-sided. In the present chapter we consider these further points of difference, between one individual, or group, and another. In other words, we shall discuss the amount of spread and the degree of lop-sidedness, or skewness.

Kinds of curve

The average is a measure of where a centre is located in any distribution; i.e. it is a *measure of location*. Although the centres, mode, median and mean are useful as clues to the values of central items, they do not tell us how the items are spread, or dispersed, throughout the distribution.

To get a more complete picture of a distribution than the simple average will give us, we need a *measure of dispersion*.

The five curves on the graph below (Fig. 10.1) are all *unimodal* (one mode), and completely *symmetrical* (one side is exactly the same, a mirror reflection, as the other). All the curves have exactly the same mean value, and, as in curves of this character, the mean, mode and median are all equal to each other. We could tell little about the five distributions if we were told, for example, that the mean value was 16.

SKEWNESS

If some of the curves were skewed to one side, the values of the mean, mode and median would differ from each other. The averages would take up different positions according to which side of the graph the curve was skewed. Figures 10.2 and 10.3 are two curves, skewed to the left- and right-hand sides of the graph, and the student should note the different positions taken up by the three averages in curves of this kind.

The modal position is, of course, determined by the peak of the curve; the position of the median is that it divides the area under curve (however skewed)

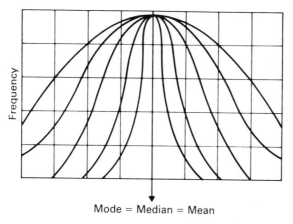

Mode = Median = Mean

Fig. 10.1

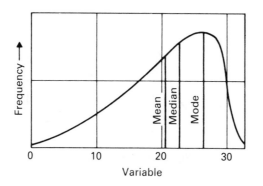

Fig. 10.2 Negatively skewed

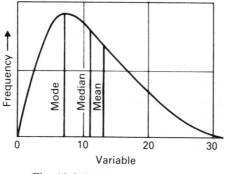

Fig. 10.3 Positively skewed

in half; the arithmetic mean is determined arithmetically and its position cannot be found merely by inspecting the curve. The more a curve is skewed away from the symmetrical, the more will the averages pull away from each other. In a skewed curve the median is always to be found between the other two averages, and, as the position of the mode is the most obvious, the student should be able to draw the approximate positions of each average on almost any curve from this simple knowledge. The student might, indeed, have a vague idea of the dispersion of a distribution if the values of mode, median and mean were given.

Among the infinite variety of ways in which a distribution may be skewed, certain types of distribution are well known to the statistician. Here, some of them are described below with brief notes on the occasions on which they are found.

The normal curve

This curve is unimodal and completely symmetrical. It is often described as having a 'cocked hat' or a 'bell' shape. The three averages are, of course, identical in this type of curve.

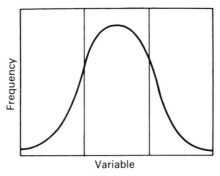

Fig. 10.4 Normal curve

Typical distributions are:

1. Intelligence of a population.
2. Height, weight, and generally in biological data.
3. Chiefly important in statistical sampling theory.

The J-shaped curve

This curve is extremely skewed (asymmetrical), and it can hardly be said to have a mode (this being cut at the zero line). The curve can be skewed either negatively or positively.

Typical positive distributions are:

1. Income of a population.
2. Property holding among a population.
3. Size of community groups (towns, villages, etc.) in a country.
4. Size of firms.

Typical negative distributions are more rarely met with, and usually applicable to select instances, e.g. numbers of offspring of animals and insects, height of land in mountainous regions, etc.

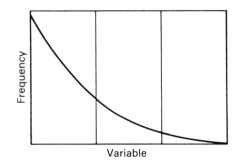

Fig. 10.5 J-shaped curve (positive)

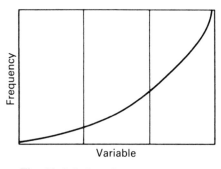

Fig. 10.6 J-shaped curve (negative)

A typical example of this type is the supply curve in economics, where the vertical scale is the price of the commodity, and the horizontal one is the quantity supplied by producers; i.e. the higher the price, the more will be produced.

The U-shaped curve
This curve is often a combination of the previous two J-shaped curves and, if the values of the variable are made very small (class intervals), it may be found to be bimodal (two modes).

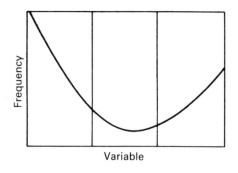

Fig. 10.7 U-shaped curve

Typical distributions are:

1. Frequency of deaths in the very young and the very old.
2. Degree of cloudiness in northern and southern temperate regions.
3. Where two different groups of items have been mixed and graphed in one curve, e.g. unskilled and skilled workers (in this case the ends of the curve might be bent over to form a bimodal curve).

The most common curves are variations or modifications of these extreme types. Many curves are combinations of one or more types. An example is given below (Fig. 10.8) of a curve which has more than one mode and may be regarded as a combination of a U-shaped curve, or two J-shaped curves (positive and negative), and a moderately symmetrical unimodal curve.

The high infantile mortality (deaths of infants up to one year of age) and the frequent death of young children are responsible for the positively skewed, J-shaped part of the curve. The moderately symmetrical unimodal curve centred on the early twenties is probably the result of accidental deaths of men and women in the more active part of their lives (e.g. sporting fatalities; much greater risk of accidents to women in the home with electrical equipment, etc.; women going out to work and meeting with accidents on the street or at work, etc.). The negatively skewed curve is the mounting curve of deaths largely from the diseases of middle and old age.

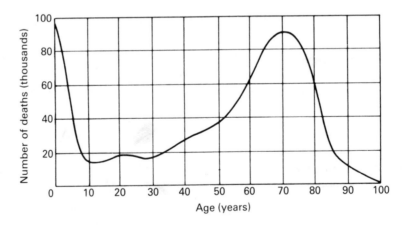

Fig. 10.8 Ages at death in a population

To summarize, so far, we have found three characteristics of any unimodal curve we care to draw:

1. Location.
2. Dispersion.
3. Skewness.

Three graphs (Figs 10.9, 10.10 and 10.11), showing clearly where these differences might occur, appear on below).

Unless we are to quote a distribution in full, or present a graph of it, we must have some short, numerical measures of dispersion and skewness. The remainder of this Method section is concerned with such measures in common use.

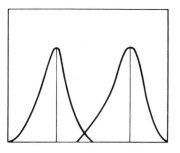

Fig. 10.9 Different location

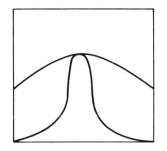

Fig. 10.10 Different dispersion

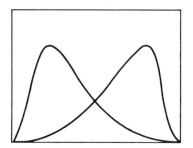

Fig. 10.11 Different skewness

The range

This is the simplest measure of dispersion and consists merely in quoting the values of the extreme items at either end of the distribution. In other words, the range is the highest and the lowest values in a distribution. It is often quoted as the actual *difference* between the two extreme values. For example, if the highest wage is £150 and the lowest is £100, then the range is £150 − £100 = £50. It has been said before that the odd, peculiarly high or low value, which is not really characteristic of the distribution but which sometimes creeps in, can distort the calculation of an otherwise worthwhile average—i.e. the arithmetic mean. Thus, in such a distribution not only the arithmetic mean but also the range will be uncharacteristic and consequently of little use. The range is the easiest measure of dispersion to understand and the easiest to obtain; it can also be the most faulty, and in any case does not tell us anything of the dispersion of values *between* the highest and the lowest.

The mean deviation

A better measure of dispersion than the range is the mean deviation which gives the average of all the deviations of items from the value of the arithmetic mean. Sometimes the average of the deviations from the median is taken, if this is more suitable for the purpose in hand.

Examples are given in Tables 10.1 and 10.2 of the calculation of the mean deviation from the arithmetic mean and from the median.

Table 10.1

Value (£s)	Frequency	Total values (products) (£s)		Deviations from mean (£s)	Deviations × frequency (£s)
0	1	0		3	3
1	2	2		2	4
2	2	4	A.M. = $\dfrac{£51}{17}$	1	2
3	3	9		0	0
4	9	36	= £3	1	9
TOTALS	17	51		—	18

$$\text{Mean deviation} = \frac{£18}{17} = £1.06$$

Table 10.2

Value (£s)	Frequency	Cumulative frequency		Deviation from median (£s)	Deviations × frequency (£s)
0	1	1		4	4
1	2	3		3	6
2	2	5	Median is 9th	2	4
3	3	8	item = £4	1	3
4	9	17		0	0
TOTALS	17	—		—	17

$$\text{Mean deviation} = \frac{£17}{17} = £1$$

It should be noted that in this calculation the signs (+ or −) of the deviations from each average are ignored because otherwise, in the case of the first calculation, we would have got a zero, meaningless, result. In any case, the mean deviation is a measure of the extent of the *average of all deviations* and to obtain this we must treat positive and negative deviations on the same footing. *Note.* The deviation from the arithmetic mean is never lower than the deviation from the median.

The quartile deviation

To understand this measure of dispersion, it it best to examine the divisions of the cumulative frequency curve which are usually made. It will be remembered from the previous chapter that a distribution can be divided by the median, so that half the number of items lie above this average and half the number below it. This is shown on the cumulative frequency curve in Fig. 10.12.

The *quartiles* are also shown on the ogive. The lower quartile is a quarter of the way along the distribution arranged in ascending order, and the upper quartile is three-quarters of the way from the lower end. From the ogive the upper quartile may be estimated at £28, the lower quartile at £13, and the median value at £21.

As these three positions divide the distribution into quarters, so can we divide it into *deciles* (tenths) and *percentiles* (hundredths). Such divisions are used extensively in intelligence testing and examination marking because they provide a more detailed picture of the dispersion than do the quartiles.

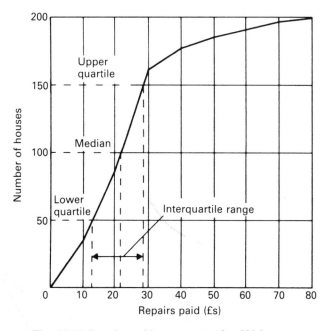

Fig. 10.12 Repairs paid per quarter for 200 houses

The quartile deviation, or *semi-interquartile range*, is found by determining the *interquartile range* (upper quartile − lower quartile) and dividing this by two. Thus, from the estimated figures in our graph:

$$\frac{£28 - £13}{2}$$ gives £7.50 as the quartile deviation.

It should be realized that this measure of dispersion includes only that half of the distribution centred on the median; therefore, it does not really cover the whole distribution. Because half the number of total items fall within the two quartiles, it follows that the smaller the quartile deviation, the more closely packed these items are, i.e. the less they are spread, or dispersed.

A defect of this measure is that the answer will be in some sort of unit—in this case £s. Consequently, it cannot be used to compare two distributions which are expressed in different units. This is also true of the other measures we have discussed.

Table 10.3 Repairs paid per quarter for 200 houses

Quarterly repairs paid (£s)	Number of houses	Cumulative frequency
0– 9.99	35	35
10–19.99	52	87
20–29.99	76	163
30–39.99	15	178
40–49.99	9	187
50–59.99	6	193
60–69.99	4	197
70–79.99	3	200
TOTAL	200	

The quartile coefficient of dispersion

To overcome this difficulty we may divide the answer by the median or, what is roughly the same thing, by half the sum of the quartiles. The whole operation may now be regarded as follows:

$$\frac{\dfrac{Q3 - Q1}{2}}{\dfrac{Q3 + Q1}{2}}$$

We can multiply both the numerator and denominator by two, which does not affect the answer, but simplifies the calculation.

This gives us:

$$\frac{Q3 - Q1}{Q3 + Q1} \quad \text{or} \quad \frac{\text{the difference between quartiles}}{\text{the sum of quartiles}} \qquad (F.10.1)$$

The units cancel out, and we get the *quartile coefficient of dispersion*. The calculation for Fig. 10.12 is:

$$\frac{£28 - £13}{£28 + £13} = \frac{£15}{£41} = 0.4$$

Note. The answer is always less than unity, since the sum of two numbers (denominator) is always greater than their difference.

The last measure of dispersion to be discussed, and the most valuable and widely used, is the standard deviation (S.D.). In the calculation of this measure of dispersion the following steps are taken: **The standard deviation**

1. Find the arithmetic mean of the distribution.
2. Find the deviations of the value of all items from the arithmetic mean.
3. Square each deviation.
4. Add the squared deviations.
5. Divide the total of squared deviations by the number of items. (The result is the *variance*.)
6. Find the square root of the variance. The result is the *standard deviation*.

A short example (Table 10.4) will serve to make this rather lengthy explanation a little clearer:

Table 10.4

Value of item		Deviations from A.M.	Square of deviations
5		−2	4
8	A.M. $= \dfrac{35}{5} = 7$	+1	1
12		+5	25
3		−4	16
7		0	0
35	TOTALS		46

$$\text{Variance} = \frac{46}{5} = 9.2$$

$$\text{Standard deviation} = \sqrt{9.2} = 3.03$$

In this example, *ungrouped data* have been used in a very short calculation. At this stage, the student is simply asked to note the steps (which, of course, are exactly the same for more lengthy calculations), and the fact that the deviations have been *squared*, in order to get rid of the plus and minus signs. The method of merely ignoring the signs, which was used in calculating the mean deviation, is really mathematically incorrect, and is one of the faults of that measure of dispersion.

For *grouped data* the calculation of the standard deviation is a little more complex. An example is given in Table 10.5.

Table 10.5 A household survey on food expenditure per week in 400 households

Column 1	Column 2	Column 3	Column 4 (2 × 3)	Column 5		Column 6 (2 × 5)		Column 7 (6 × 5)
Expenditure in £s	Number of households	Midpoints	Frequency × midpoints	Deviation of column 3 from arithmetic mean		Frequency × deviation		Frequency × deviation^2
				−	+	−	+	
55 and under 65	10	60	600	40		400		16 000
65 and under 75	14	70	980	30		420		12 600
75 and under 85	45	80	3 600	20		900		18 000
85 and under 95	70	90	6 300	10		700		7 000
95 and under 105	105	100	10 500	—	—	—	—	—
105 and under 115	92	110	10 120		10		920	9 200
115 and under 125	42	120	5 040		20		840	16 800
125 and under 135	22	130	2 860		30		660	19 800
TOTALS	400	—	40 000		—	2420	2420	99 400

$$\text{Arithmetic mean} = £\,\frac{40\ 000}{400} = £100$$

$$\text{Variance} = £\,\frac{99\ 400}{400} = £284.5$$

$$\text{Standard deviation} = \sqrt{248.5} = £15.77$$

As in the above example, where class intervals include a range of values, the midpoint of the classes is found. Thus, up to and including the fourth column of our example we have simply calculated the arithmetic mean as was explained in the last chapter. The standard deviation is calculated from the last three columns only.

The deviations from the arithmetic mean, column 5, are each multiplied by the corresponding frequencies to form column 6. In the last column are the frequency × squared deviations, and in order to make this last calculation simpler, simply multiply column 6 by column 5, i.e. frequency × deviation × deviation (again) = frequency × deviation².

This is the 'long method' of calculating the standard deviation, and, had the arithmetic mean run to decimal places, the calculation would have been extremely laborious. Luckily, the arithmetic mean worked out as a whole number, and because it ended in 0, the calculation was made even simpler! In fact, with a distribution of any considerable length, this 'long method' is never used for working out the standard deviation.

SHORT METHOD

In Chapter 9, we described a method of calculating the arithmetic mean from grouped data by using an assumed mean, and correcting our answer accordingly (see Table 9.7).

We now extend this method to include the standard deviation, taking the example of the present chapter (Table 10.5).

Table 10.6 Expenditure on food for 400 households

Column 1	Column 2	Column 3	Column 4		Column 5 (2 × 4)		Column 6 (4 × 5)
Expenditure in £s	Number of households	Midpoints	Deviation from assumed A.M. of £90 units £10		Frequency × deviation		Frequency × deviation²
			−	+	−	+	
55 and under 65	10	60	3		30		90
65 and under 75	14	70	2		28		56
75 and under 85	45	80	1		45		45
85 and under 95	70	90	0		—	—	0
95 and under 105	105	100		1		105	105
105 and under 115	92	110		2		184	368
115 and under 125	42	120		3		126	378
125 and under 135	22	130		4		88	352
TOTALS	400	—	—	—	103	503	1394
						−103	
						+400	

$$\text{Variance} = \left[\frac{1394}{400} - \left(+\frac{400}{400} \right)^2 \right] \text{ in units of £10}$$

$$= 3.485 - 1^2 = 3.485 - 1 = 2.485 \text{ units of £10}$$

$$\text{Standard deviation} = \sqrt{2.485} = 1.577 \text{ units}$$

Since the answer is in units of £10, we now convert it to £s:

$$\text{Standard deviation} = 1.577 \times £10 = £15.77$$

The first points to note here are that, not only are the figures much simpler, but also one column has been dropped from the table. The deviations from the assumed mean have been counted as -4, -3, -2, etc., and, as these have eventually to be squared, this cuts out unwieldy figures in the last two columns. The student must remember that, when doing this, the result is in units of 'class intervals', and that in this example the result must be *reconverted* to £s by multiplying by £10. As we work, for simplicity, on an assumed mean (this is assumed as a convenient midpoint, usually of the largest class near the middle of the distribution), we must apply a correction when we calculate the variance. This correction is column 5 divided by the number of items, then squared, i.e.:

$$\left(\frac{\text{Frequencies} \times \text{deviations}}{\text{Number of items}}\right)^2 = \left[\frac{(+503) + (-103)}{400}\right]^2$$

Note. The student must beware of the common mistake of using the last column 'frequency $\times$ deviations^2' in the correction.

The correction must be worked out (including squaring), then subtracted from the main fraction *before* the final square root is taken. In our example, the correction was unity (because, as was known previously, the true mean was actually 100, one class interval away).

Every attempt should be made to memorize and understand the columns, layout and principles of this short method, because in most cases it is the only sensible method to use for cutting out much hard work unless your calculator is programmed to perform this operation.

The student should particularly note the following points:

1. Always work in units equal to the class interval (in this case £10).
2. Always choose, as the assumed mean, one of the midpoints. In theory, any one will do but, in practice, the working is simplified by picking one near the middle of the table.
3. The correction is *always* subtracted from the assumed variance, whether the total of column 5 is positive or negative.
4. Leave the conversion into actual units until the final result is obtained, but *do not forget to do it.*
5. Examination questions often call for the calculation of both the arithmetic mean and standard deviation from the same table. This method produces both answers from one set of calculations.

$$\text{A.M.} = £90 + \left(\frac{400}{400} \times £10\right)$$

$$= £90 + £10 = £100$$

Note. For the arithmetic mean, the correction is multiplied by the class interval (units of £10) *before* it is added to (or subtracted from) the assumed mean. In the case of the arithmetic mean, rule (3) above does not apply. If the balancing figure of column 6 is *negative* then the fraction is *subtracted* from the assumed mean.

To obtain a numerical measure of skewness, we rely on the fact that the greater this is, the more the averages are pulled apart. In Fig. 10.2, the three averages are: **Measuring skewness**

$$\text{Mean} = 20\tfrac{1}{2} \qquad \text{Median} = 22\tfrac{1}{2} \qquad \text{Mode} = 26\tfrac{1}{2}$$

Roughly speaking, the median is two-thirds of the distance between the mean and the mode, measuring from the mode.

For example, the difference in this case is $26\tfrac{1}{2} - 20\tfrac{1}{2} = 6$. Two-thirds of this is 4 and this, when subtracted from the modal value of $26\tfrac{1}{2}$, gives an estimated median value of $22\tfrac{1}{2}$.

We can use this information in two ways:

1. Subtract the mode from the arithmetic mean and divide the answer by the standard deviation.
2. Take three times the difference between the arithmetic mean and the median, i.e $3(\text{A.M.} - \text{median})$, and divide the result by the standard deviation.

If the distribution has *positive* skewness, the mode is smaller than the median which, in turn, is smaller than the arithmetic mean, and the answer will in each case be positive. If there is *negative* skewness, the arithmetic mean and median are both smaller than the mode, so the result of the subtraction is negative. These measures thus indicate the *kind* of skewness as well as its extent. The reason for dividing by the standard deviation is to get rid of the units (they cancel out) so that distributions with different units can be compared.

Use and interpretation

We have now dealt with four measures of dispersion:

1. The range.
2. The mean deviation.
3. The quartile deviation.
4. The standard deviation.

Among these measures, the standard deviation is by far the most important in use, although the desirable qualities in a measure of dispersion are not found completely in the standard deviation. Such desirable qualities are very similar to the desirable qualities for an average (see Chapter 9). They include ease of calculation and ease of understanding, stability when extreme items are involved, coverage of all the items in the distribution, and value in use and in advanced work.

The standard deviation is also the most troublesome to calculate and not the easiest to comprehend—the other three are probably far superior in these respects. Both the mean deviation and the standard deviation cover all the items in the distribution, while the other two do not. The range is the worst affected by extreme items; the quartile deviation is not affected at all, and the use of the standard deviation offsets them to some degree. The superiority of the standard deviation to all the others is in its flexibility in use to solve many problems in advanced statistics, and particularly its wide use in sampling theory.

A good deal has already been said about the first three measures. The range has a limited use in the field of quality control; the mean deviation is occasionally used in economics and social statistics; and the quartile deviation is used in educational statistics quite extensively. However, the standard deviation is used far more than the others put together, and the remainder of this chapter is largely devoted to describing this.

Relations between the measures

The last worked examples concerned the expenditure on food of 400 households. The standard deviation was calculated as £15.77, approximately, and the arithmetic mean as £100. The range is approximately £80 (i.e. £135–£55). The mean deviation can be readily worked out from Table 10.5 (long method) as:

$$\frac{\text{Total of frequencies} \times \text{deviations (ignoring signs)}}{\text{Number of items}} \quad \text{(F.10.2)}$$

$$= \frac{2420 \times 2420}{400} = \frac{4840}{400} = £12.1$$

The median and the quartile deviation are estimated from the ogive shown in Fig. 10.13, from which it can be seen that the median is approximately £101. *Note.* We should expect this to be similar to the arithmetic mean in a fairly symmetrical distribution as is the case here.

The quartile deviation is, of course, found by:

$$\frac{\text{Upper quartile} - \text{lower quartile}}{2} = \frac{£111 - £89}{2} = £11$$

The mode can be estimated from the method given in Chapter 9. The modal class is obviously '95 and under 105', and as the number of frequencies (92) is larger in the class above than in the class below (70), then the mode is slightly nearer 105 than 95. Subtracting the two frequencies on either side of the modal class from the modal frequency itself we get:

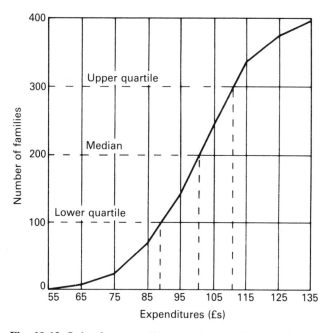

Fig. 10.13 Ogive for expenditure on food of 400 households

$$105 - 70 = 35$$

$$\text{and } 105 - 92 = 13$$

The extent of the pull from £95 (lower limit) is:

$$\text{Mode} = £95 + \frac{35}{35 + 13} \times £10 = £95 + £7.3 = £102.3$$

A list can be made of the averages and dispersion measures for this distribution:

Mode = £102.3
Arithmetic mean = £100
Median = £101
Range = £80
Mean deviation = £12.1
Quartile deviation = £11
Standard deviation = £16 (approximately)

For any distribution which is *symmetrical* (*or fairly symmetrical*) and also *unimodal*, certain relationships have been calculated between these measures.

143

1. *The values of the averages are similar.*
   (As can be seen, the averages are very similar, the only slight difference being in the case of the mode (£102.3) which, in any case, is an estimated figure.)
2. *The range should be approximately equal to six standard deviations.*
   (Six standard deviations are 6 × £15.77 = £95, and the range is £80.)
3. *The quartile deviation is approximately equal to two-thirds of the standard deviation.*
   (Two-thirds of the standard deviation is $\frac{2}{3}$ × £15.77 = £10.51. The quartile deviation is £11.)
4. *The mean deviation is approximately four-fifths of the standard deviation.*
   (Four-fifths of the standard deviation is $\frac{4}{5}$ × £15.77 = £12.62. The mean deviation is a little lower than this—£12.1.)

From our list, then, without knowing any of the actual values of the items we can tell that the curve is: *fairly symmetrical with only a very slight negative skewness (i.e. mode a little higher than median), and it is unimodal, centred on £100, with most of the items lying within £48 either side of this figure.*

Nature and meaning of the standard deviation

It is not sufficient to know merely how to calculate the standard deviation; the student must know its meaning. This is quite simple and is shown in Fig. 10.14 below.

This curve is unimodal and symmetrical about the average; in fact, it is of the 'normal curve' type described in Fig. 10.4. On either side of the mean are marked off three standard deviations—making six standard deviations for

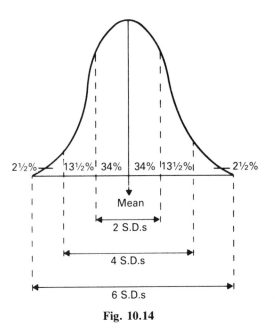

Fig. 10.14

approximately the whole range of values. It has been calculated that if we mark off a frequency curve in this way, the area under the curve (which is proportional to the total number of items) will be divided into certain proportions. The percentage figures under the curve give these approximate proportions:

- Within one standard deviation either side of the mean (two standard deviations in all) will lie approximately 68 per cent of the items.
- Within two standard deviations either side of the mean (four standard deviations in all) will lie approximately 95 per cent of the items.
- Within three standard deviations either side of the mean (six standard deviations in all) will lie approximately 100 per cent of the items.

The student will have realized by now that all measures of dispersion and location form a shorthand for giving us a picture of distributions quickly, rather than by building up a picture by other tedious working.

It will be remembered that we built up a picture of the curve (see description following Fig. 10.13) from the list of average and dispersion measures of 'Expenditure on food for 400 households'. Actually, there was no need to calculate all these measures, although we did this to establish the relationships. All that is necessary to get a fairly true picture of the curve is to know the mean and the standard deviation, and the number of items.

Mean = £100
Standard deviation = £16
Number of items = 400

With this simple information we can:

1. Draw a horizontal axis and mark off £100 at the middle.
2. Erect a vertical at this £100 mark and label this the mean.
3. From this £100 on the horizontal axis mark off to each side three standard deviations and scale them in £s. For example, to the right of the mean will be:
   £116 = (£100 + £16)
   £132 = (£100 + £16 + £16)
   £148 = (£100 + £16 + £16 + £16)
   and the lower values for the left-hand side.
4. Calculate the number of items (from the percentages) which will make up the six divisions of the curve (standard deviations) and mark in.
5. Remembering that the number of items within each section is proportional to the area of each section, it will now be possible to draw a rough curve to represent the distribution. This is done shown in Fig. 10.15, and the figures entered could be compared by the student with the actual figures of the distribution on page 146.

Such a figure need not be drawn but with a little practice could be imagined or, at the most, sketched rapidly.

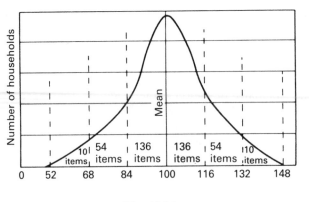

Fig. 10.15

Some applications	The standard deviation may be used as a measure of dispersion in all symmetrical, unimodal and even moderately skewed distributions. Such distributions are frequently met with, and they include intelligence testing, examination marking and many kinds of natural and biological series. In sampling theory, this measure is of great importance to the statistician, but its applications in this respect are considered in later chapters.
Coefficient of variation	Finally, the standard deviation can be used to compare the dispersion in two distributions as long as the distributions are of the same kind and are measured in the same units. For this purpose we use the *coefficient of variation*. To calculate this, we simply multiply the standard deviation by 100 to bring it to a percentage, and divide by the arithmetic mean, for each distribution.

Example:

Coal output per person over a period at two collieries was measured as follows:

Mine A   A.M. 45 000 kg, S.D. 15 000 kg
Mine B   A.M. 40 000 kg, S.D. 14 000 kg

applying the formula—

$$\text{Coefficient of variation} = \frac{\text{standard deviation}}{\text{arithmetic mean}} \times 100 \qquad (\text{F.10.3})$$

we get:

Mine A $\dfrac{15}{45} \times 100 = 33.33$ per cent

Mine B $\dfrac{14}{40} \times 100 = 35.00$ per cent

146

The conclusion may be drawn that, although Mine A has a higher mean output, and although the standard deviation might suggest that the values of output were more dispersed than in Mine B, nevertheless, the dispersion, relative to the mean, is higher in Mine B, as the coefficient of variation shows.

Taking the example of Table 10.5, we have already found the following **Skewness** measures:

Arithmetic mean = £100
Median = £101
Mode = £102.3 (say £102)
Standard deviation = £16 (approximately)

Using *Method 1.* (based on the mode) we have:

$$\frac{\text{A.M.} - \text{mode}}{\text{S.D.}} = \frac{£100 - £102}{£16} = -\frac{2}{16} = -0.125$$

If the mode is not well defined (as often happens when the frequency curve has a plateau rather than a peak) we use *Method 2.* (based on the median). We have:

$$\frac{3(\text{A.M.} - \text{median})}{\text{S.D.}} = \frac{3(£100 - £101)}{£16} = -\frac{3}{16} = -0.188$$

Note. The two methods do not necessarily give the same answer, and in any case we have approximated both the mode and the S.D.

The result indicates a moderate degree of negative skewness. A highly skewed distribution may have a value of ±1, and values up to ±3 are theoretically possible.

IMPORTANCE OF SKEWNESS

This notion is used in certain theoretical distributions (see Chapter 14 on Sampling).

It is also useful in industrial and economic statistics to know, for example, that the distribution of incomes is skewed, and to what extent. An employer would find that the age distribution of his workers was skewed, and probably their earnings under a bonus or productivity scheme. The interesting question would then arise, as to whether these two cases of skewness differed at all, in nature or extent. That is, were the older workers also the ones who earned most bonus, or was it the reverse?

Dispersion and computer packages

MINITAB commands

Averages and measures of dispersion are easily obtained in MINITAB. Commands like MEAN C1, RANGE 4, and STD C2 (for standard deviation) produce individual statistics. A useful command is DESC (short for DESCribe). DESC produces a variety of descriptive statistics for a variable. The command can call up these statistics for more than one variable, e.g.:

$$MTB > DESC\ C1-C4$$

First steps with SPSS^x

The statistical package **SPSS** (Statistical Package for the Social Sciences) is a powerful tool of analysis, developed over more than a quarter of a century. Its modern version for use on large mainframe computers is **SPSS^x**. Most users of SPSS^x know it as a non-interactive package—i.e. the results of each command do not normally appear instantaneously on the screen once ⟨**RETURN**⟩ or ⟨**Enter**⟩ has been pressed. See Chapter 4 for a general discussion of the characteristics of interactive and non-interactive working.

Users of **SPSS/PC+** should read the present chapter. It gives a flavour of SPSS software before SPSS/PC+ is directly met in Chapter 11.

We want to use SPSS^x to calculate several measures of central tendency (i.e. averages) and measures of dispersion for the products data in Table 10.1 above. In this example the data will be input along with the SPSS^x commands that define the data and produce statistics. Imagine that we have, for the 17 articles or 'cases', information on product type and colour, as well as value (variable PRICE). We therefore have three variables and 17 cases.

SPSS^x commands and data are input into a 'command file' (also known as 'job file' or 'run file'. This file is created and filled outside the SPSS^x package using the *editor* software installed on the mainframe computer. The editor can be thought of as a very simple wordprocessor. The procedures for calling up and using the editor will vary between computers. Using the editor we input the following combination of commands and data. The command file is stored under a name that we choose, here 'RUN1'. Each line is followed by ⟨**RETURN**⟩ or ⟨**Enter**⟩.

```
TITLE              RUN1: PRODUCT DATA ANALYSIS with SPSSˣ
DATA LIST          FILE = INLINE
                   / PRICE 1, C2 TO C3 2-5
VARIABLE LABELS    C2 'TYPE'/ C3 'COLOUR'
VALUE LABELS       C2 1 'BOOKLET', 2 'SPARE', 3 'ACCESSORY' /
                   C3 1 'RED', 2 'WHITE', 3 'BLUE'
MISSING VALUES     C2 TO C3 (9)
BEGIN DATA
0 1 9
1 1 1
        . . .
        . . .
        . . .
4 3 1
4 1 2
END DATA
```

```
FREQUENCIES          VARIABLES = PRICE /
                     STATISTICS = DEFAULT MEDIAN MODE RANGE
                     SKEWNESS
FINISH
```

This SPSS^x run turns out to be less complex than it may seem at first sight. The TITLE command is a simple cosmetic convenience and is optional. DATA LIST, however, is necessary. The detail that follows (FILE = INLINE) will tell the package that the data is to be input along with the commands, i.e. is 'inline'. The second line of DATA LIST declares that there will be three variables for each case—'PRICE' (value of the product) in column 1, plus 'C2' and 'C3' occupying columns 2 to 5 (inclusive). SPSS^x will understand that the variables C2 and C3 will each occupy two columns (2–3, 4–5). In the jargon of SPSS^x the details of the variable names, and which columns they occupy, are *specifications*. The VARIABLE LABELS command labels C2 and C3 as 'TYPE' and 'COLOUR' respectively. VALUE LABELS gives labels to the values of these two variables. Both these label commands are optional, and both normally require labels to be typed within single quotation marks. Also optional, but usually advisable, is the MISSING VALUES command. Here the package is instructed to treat the value '9' as missing data for both variables C2 and C3. These last three lines all illustrate how SPSS^x command lines are usually built from *command keywords* followed by *specifications*.

This concludes the data definition section of the command file. The data is introduced after the BEGIN DATA command (without further specification). The data actually begins on the next line. In this text we have omitted all but 4 of the 17 data cases. The command END DATA lets the package know that data entry has finished.

The command that actually produces data analysis in this example is the SPSS^x statistical *procedure* FREQUENCIES. FREQUENCIES gives a simple frequency table of all the values of the variable(s) declared in the VARIABLES subcommand, here PRICE. As part of the FREQUENCIES procedure we have also asked for some descriptive STATISTICS. There is a bundle of 'DEFAULT' statistics (mean, standard deviation, minimum, maximum) plus four others. The command file ends with a FINISH command.

Note that SPSS^x grammar uses the forward slash ('/') to mark divisions between various parts of the declarations. The commas are optional—a space would have been enough. We have included the commas to make the command file more readable. In general, SPSS^x grammar is fairly forgiving (e.g. in accepting various spacings between the end of the command keywords and the specifications or declarations). Note, too, that in the lines of data the SPSS^x package accepts only one digit per column (compare MINITAB). In lines of data input to SPSS^x, spaces are treated as columns—that is why in the DATA LIST command the dataset is declared as occupying five columns although there are only three variables (PRICE, C2, C3). Also note the SPSS^x keywords such as 'T0' and the symbols like '=' and '/'. Since C2 and C3 are not used in the analysis all references to them in RUN1 are really superfluous. C2 and C3 are included here solely to illustrate the use of SPSS^x commands and grammar.

Users of SPSS/PC+ will find that the SPSS/PC+ commands FREQUENCIES and DESCRIPTIVES (see Chapter 11) closely resemble the SPSS^x procedures introduced here.

It is easy to become overawed by the need to get the grammar of an SPSS^x command file correct. The excellent reference manuals published by SPSS^x may also be intimidating. But do not worry! It is not really that complicated. The package is best learned by following examples, like the one above. Just bear in mind a few guidelines:

- **Command keywords** like FREQUENCIES must begin in *column 1*.
- Words within **command phrases** (e.g. DATA LIST) are separated by a *single space*.
- **Command lines** (and lines of data) should not normally be more than *80 columns* long, but may be continued.
- **Continuation command lines** must *not* begin in column 1.

There are other rules of SPSS^x grammar, but they become rather tedious to list. The best strategy is to lay out your own command files like the one above—they will be both readable (to the human eye) and generally acceptable to the SPSS^x package.

A FIRST SPSS^x RUN

Once we have keyed the commands and data into the command file we save the command file under our chosen name, here 'RUN1', and we exit from the computer's editor software. Leaving the editor does not automatically activate SPSS^x. SPSS^x has to be called explicitly. The exact form of the operating system command to do this will vary between mainframe computers. On a PRIME computer the command is typically as shown below. Follow the command with ⟨**RETURN**⟩.

```
spss^x -i run1 -1 out1
```

First SPSS^x is called. The tag '−i' indicates that the commands are to be found in input (command) file RUN1. The '−1' tag tells the system to direct output to an output 'listing' file (compare MINITAB's OUTFILE command) called 'OUT1' (also a file name chosen by ourselves).

The computer answers with various messages. If the run fails we shall see an error message. In that case the RUN1 command file will have to be accessed again with the editor software and the grammar of the commands amended. SPSS^x will then be recalled (as in the command above, or variant for your computer). When the run is successful we are told that the command file has been processed without errors of grammar being detected. In this style of operation we do not normally automatically see the results on the screen. They have been stored in the output file OUT1. This file can be displayed on the screen using the appropriate operating system command for your computer, e.g.

150

FLIST RUN1 (for a PRIME machine). On screen, the output text may seem a little confused. Do not worry; this is probably because the lines of OUT1 are too long to fit comfortably on the standard VDU screen. The output will appear clearer when OUT1 is directed to an appropriate paper printer.

SPSS^x output is quite extensive. There will be various starting messages, plus a repetition of the commands contained in the command file RUN1. Fig. 10.16 below reproduces the most important section of OUT1—the results of the FREQUENCIES procedure:

PRICE

VALUE	FREQUENCY	PERCENT	VALID PERCENT	CUM PERCENT
0	1	5.9	5.9	5.9
1	2	11.8	11.8	17.6
2	2	11.8	11.8	29.4
3	3	17.6	17.6	47.1
4	9	52.9	52.9	100.0
TOTAL	17	100.0	100.0	

MEAN	3.000	MEDIAN	4.000	MODE	4.000
STD DEV	1.323	SKEWNESS	−1.101	RANGE	4.000
MINIMUM	.000	MAXIMUM	4.000		
VALID CASES	17	MISSING CASES	0		

Fig. 10.16 SPSS^x output from file OUT1 (product data analysis)

Note that because there are no missing cases (i.e. there is valid data on PRICE for all 17 cases) the 'Percent' and 'Valid percent' columns are identical.

A SECOND SPSS^x EXAMPLE

In surveys of even moderate size it is practically unsatisfactory to include data in the command file every time that we wish to investigate some aspect of our dataset. As with our brief introductory career with MINITAB (see Chapters 5–8), we must soon learn how to read data into SPSS^x procedures from a separate stored data file.

In this second SPSS^x example we recalculate the mean and standard deviation for the household food expenditure data in Table 10.5. We have input the data, using the mainframe computer's editor, to a file called 'HOUSE'. You will notice that in Table 10.5 the data for the 400 households have already been 'reduced' to a grouped frequency table. The original data, as input to a computer data file, would almost certainly be recorded in a 'raw' state—i.e. the actual amount (to the nearest £) spent by each household. Our analysis below will use actual amounts, rather than group the expenditure data into the eight groups or classes of Table 10.5. The power of computers is such that data reduction into broad classes is not always so necessary as with 'hand and calculator' calculation.

Since surveys rarely collect just one piece of information per unit, we have made our example dataset more realistic by including a second variable, region (i.e. location of the household), in five categories. The first few lines of our data file HOUSE look as follows:

```
072 3
124 1
093 4
133 1
```

We have information on 400 households. In SPSS^x terminology we have 400 cases. In the HOUSE file we have very little information per household—the data for each household occupies only a few columns per line or *record*. The maximum normal record length for an SPSS^x data file is *80 columns*. However, where we have lots of information about each case, data may occupy more than one record. In larger datasets like this, the DATA LIST command has to be more complex.

The editor has been used to create and store the data file HOUSE. It is also used to create an SPSS^x command file, 'RUN2'. The file RUN2 calls data from HOUSE and analyses this data. The SPSS^x statistical procedure is CONDESCRIPTIVE. It is like FREQUENCIES, but ideally suited to interval level and ratio level variables (see Chapter 3) that can take many values. With CONDESCRIPTIVE there is no frequency table, just appropriate statistics. The RUN2 file is reproduced below:

```
TITLE              RUN2: HOUSEHOLD FOOD EXPENDITURE with SPSS^x
FILE HANDLE        INDATA / NAME = 'HOUSE'
DATA LIST          FILE = INDATA
                   / EXPEND 1-3, C2 4-5
VARIABLE LABELS    EXPEND 'EXPENDITURE IN POUNDS' / C2 'REGION'
VALUE LABELS       C2 1 'N. IRELAND', 2 'WALES', 3 'SCOTLAND'
                   4 'N. ENGLAND', 5 'S. ENGLAND' /
MISSING VALUES     EXPEND (999) / C2 (9)
CONDESCRIPTIVE     EXPEND
STATISTICS         1 5 8 10 11
FINISH
```

The RUN2 command file introduces us to the FILE HANDLE command. This is used to define input and output data files in SPSS^x. Here the data file that contributes the data to the analysis is introduced with NAME = 'HOUSE' (grammar may vary between installations). HOUSE is the name of the data file created outside SPSS^x with the editor. In order to cope with the file handling necessities of certain makes of computer, SPSS^x insists that we give HOUSE a special second name by which it will be known internally within the command file. This may seem odd and confusing—but do not worry. Most humans answer to more than one name! Think of the second internal name as a nickname. Here we have chosen the filename 'INDATA' as a temporary 'nickname' for HOUSE.

RUN2 continues with the DATA LIST command in which the specified input data file (INDATA) must match the nickname specified in the FILE HANDLE line. DATA LIST also declares the two variables, 'EXPEND' and 'C2', plus their column locations. The two variables are further defined in VARIABLE LABELS, VALUE LABELS and MISSING VALUES (as in RUN1). Note where the forward slash and the single quotation marks occur in the grammar. No columns of data are included in the command file—the data will all be called from the 400 lines of the HOUSE data file.

The CONDESCRIPTIVE procedure specifies EXPEND, short for expenditure, as the variable of interest. We ask for five different statistics to be calculated. The CONDESCRIPTIVE procedure specifies statistics by a reference number (1, 5 . . .). The SPSS[x] reference manuals indicate which statistics correspond to the numbers. Here we have called for the mean, standard deviation, skewness, minimum and maximum. Note that the grammar for CONDESCRIPTIVE and its statistics differs from the grammar of the FREQUENCIES procedure in RUN1.

A SECOND ANALYSIS WITH SPSS[x]

As with the first SPSS[x] example, RUN2 is set to work (once we have exited from the editor) with a command something like:

spss[x] −i run2 −1 out2

We know that all is well if we receive the operating system message that RUN2 has been processed with no errors. Below, in a slightly truncated form, the output produced by RUN2 and stored in OUT2 is reproduced in Fig. 10.17. The output listing file of results, 'OUT2', can be printed on screen and/or paper:

```
NUMBER OF VALID OBSERVATIONS (LISTWISE) = 400.00
VARIABLE      MEAN    STD DEV  SKEWNESS   MIN    MAX    VALID N
EXPEND      99.997    15.812    −.283     55     134         400
```

Fig. 10.17 SPSS[x] output from file OUT2 (household food expenditure)

Compare the SPSS[x] results with those obtained from the grouped frequency Table 10.5. Our SPSS[x] standard deviation is slightly higher, and the mean is not quite exactly £100. These differences are to be expected. Remember that in the SPSS[x] example here, we work with the original data. Table 10.5 has already half-processed the data into groups or classes. This is convenient, but inevitably loses some detail from the data.

SPSS[x] users who need to manipulate or 'transform' data (COMPUTE, IF, RECODE) or to select data (SELECT IF) will find examples of these procedures in the final sections of Chapter 15.

11

Correlation and regression

Method

We saw in Chapter 6 that it was possible to show the existence of correlation between two sets of data by means of a scatter diagram.

While such a device will indicate the *type* of correlation (i.e. positive or negative, linear or curvi-linear) it gives no indication of its *extent*. In this chapter we consider numerical measures of correlation.

Regression lines

When there is strong correlation in a scatter diagram, the dots tend to arrange themselves in a narrow band, which may be curved or straight.

If the band of dots is straight, the correlation is said to be *linear*, i.e. the relationship between the two variables can be represented by a straight line.

If the relationship is only approximately linear, then it is possible to construct a *line of best fit*, which runs through most of the dots, and leaves the remaining ones more or less equally disposed on either side of this line. Such a line may be fitted to a scatter diagram by inspection, and this has been done in Figs 6.9, 6.10 and 6.11.

Three-point method for regression lines

This is the simplest method of calculating a regression line, but it does not yield an equation directly, as does the method which follows this section. The procedure is as follows:

1. List the two variables in vertical columns, with one column in ascending order, and find the simple average (arithmetic mean) of each column. The two values obtained, when plotted on our scatter diagram, will yield the *first* of our three points.
2. Mark the position of the arithmetic mean in the ranked column of figures. This will probably fall at some intermediate position between two of the original figures, since the arithmetic mean may not be a whole number.
3. For each column find the average (arithmetic mean) of all those figures which are positioned *above* the point marked. These two values will give our *second* point.

4. Do the same for all values *below* the point marked. This gives our *third* point.
5. The three points calculated can now be plotted on the original diagram, and when joined will form a straight line.

Example 1:
Using the figures from Table 11.1 on page 158, we have:

Sales of ice-cream (£ hundreds) Y	Midday temperature (°C) X
45	24
50	28
→	←
60	32
65	36
65	40
5)285	5)160
57	32

1. Our first point is $Y = 57$, $X = 32$.
2. We count the 32 in the X column as *below* our mark, so that we have three values in each column.
3. Our second point is $Y = \frac{1}{2}(45 + 50) = 47\frac{1}{2}$ and $X = \frac{1}{2}(24 + 28) = 26$.
4. Our third point is $Y = \frac{1}{3}(60 + 65 + 65) = 63\frac{1}{3}$ and $X = \frac{1}{3}(32 + 36 + 40) = 36$.

These three points, when plotted on our scatter diagram, will fall on a straight line.

Note. This line will not necessarily coincide with either of those obtained from the formulae, except that the *first* point, based on the arithmetic means, will lie on all three of them.

If we wish to calculate the mathematical equation for such a line, we can approach the problem in two ways:

1. We can take fixed values for the Xs (the variable on the horizontal scale of our graph) and find a line which minimizes the *vertical* distances between the dots and this line.

   This is known as the *regression line of Y upon X*. It gives us an estimate of Y (the variable shown on the vertical scale) for a known value of X.
2. We take fixed values for the Ys, and find a line which minimizes the *horizontal* distances between the dots and the line. This is called the *regression line of X upon Y*, and will give us an estimate of X from a known value of Y.

Actually, since the deviations may be positive or negative, with respect to the line, we use a method which minimizes the *squared deviations*.

For example, suppose we have two sets of figures, one for sales of ice-cream on certain days, and the other for the midday temperatures on the same days. We should expect some correlation to exist between the two, but it would not be perfect, since other factors enter into ice-cream sales, apart from temperature.

If we plot the temperatures on the horizontal scale, these would be our Xs. The sales would then be our Ys.

- Line 1 (regression of Y upon X) would give us an *estimate* of sales for a given temperature;
- Line 2 (regression of X upon Y) would give an *estimate* of temperature from a known sales figure.

The line which one would fit by inspection would be one which is half-way between the two. This would be an over-simplification of the true situation, and its position would depend upon the judgement of the individual.

The equation of a straight line

The mathematical equation for *any* straight line is:

$$Y = a + bX \qquad \text{(F.11.1)}$$

where a and b are constants for any one line.

The value of a determines where the line will cut the Y-axis on our graph, and the value of b gives the slope of the line. These values need not be whole numbers, and they can be either positive or negative.

Figures 11.1 and 11.2 show the appearance of our line for different values of a and b.

Clearly, if we know exactly where a line cuts the Y-axis, and we also know its slope, then there is only one line which satisfies these requirements; i.e. we can identify *any particular* line if we know the values of a and b.

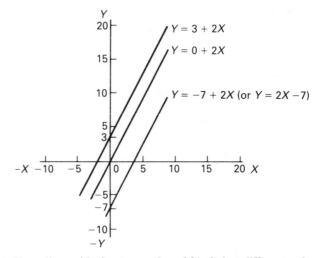

Fig. 11.1 Three lines with the same value of $b(=2)$ but different values of a

To find the values of a and b in any particular case we solve the following equations:

Regression of Y upon X

$$\Sigma Y = na + b\Sigma X \qquad \text{(F.11.2)}$$

$$\Sigma XY = a\Sigma X + b\Sigma X^2 \qquad \text{(F.11.3)}$$

where

    n = the number of *pairs* of figures.
    ΣY = the sum of the Y figures.
    ΣX = the sum of the X figures.
    ΣXY = the sum of the products of each X multiplied by the corresponding Y.
    ΣX^2 = the sum of the squares of each individual X

Note 1. Σ = the capital letter 'sigma' in the Greek alphabet. This is a shorthand method of writing 'the sum of'.

Note 2. So far as the formulae are concerned, it does not matter which set of figures we call the X series, and which the Y series. The solutions we get, however, give the values of a and b in the equation:

$$Y = a + bX$$

i.e. the line which estimates the value of Y in any particular case.

The student should therefore make sure that he or she chooses as the Y series the particular variable asked for in the question.

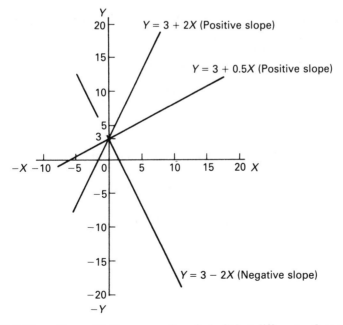

Fig. 11.2 Three lines with the same value of $a(=3)$ but different values of b

Example 2:

The following table gives ice-cream sales (in £ hundreds) on five days of the week together with the corresponding midday temperature, on the same five days. Obtain the equation to the regression line of ice-cream sales on temperature.

Note 1. Since we are asked for the equation which estimates *ice-cream sales*, we make this our Y. In any event, since this is the *dependent variable*, it would represent the vertical scale on a scatter diagram (= Y-axis).

Table 11.1

Sales of ice-cream in (£ hundreds) Y	Midday temp. (°C) X	XY	Y^2	X^2
65	40	2 600	4 225	1 600
45	24	1 080	2 025	576
65	36	2 340	4 225	1 296
50	28	1 400	2 500	784
60	32	1 920	3 600	1 024
285 ($= \Sigma Y$)	160 ($= \Sigma X$)	9 340 ($= \Sigma XY$)	16 575 ($= \Sigma XY^2$)	5 280 ($= \Sigma X^2$)

Note 2. We have added a column for ΣY^2, since this is required for the next part of the exercise.

Substituting the values for ΣY, ΣX, ΣXY and ΣX^2 in our two formulae, F11.2, F11.3, and noting that $n = 5$, we have:

$$285 = 5a + 160b \qquad \text{Eq.(11.1)}$$

$$9340 = 160a + 5280b \qquad \text{Eq.(11.2)}$$

We solve these two equations as follows:

1. Multiply Eq. 11.1 by 32 (to make the *a* value equal to Eq. 11.2)

$$9120 = 160a + 5120b \qquad \text{Eq.(11.3)}$$

2. Subtract Eq. 11.3 from Eq. 11.2 (to eliminate *a*)

$$220 = 160b$$

$$\frac{220}{160} = b \quad \text{i.e. } b = 1375$$

3. Substitute this value in Eq. 11.1

$$285 = 5a + 220$$
$$285 - 220 = 5a$$
$$65 = 5a \quad \text{i.e. } a = 13$$

The regression line is therefore:

$$Y = 13 + 1.375X \qquad \text{Eq.(11.4)}$$

We can test this, as an estimate of Y, against the original data. For example, when $X = 40$, the regression equation gives:

$$Y = 13 + 55 = 68 \ (\pounds \text{ hundreds})$$

Actually, the sales on that day were 65 ($\pounds$ hundreds), showing that part of the sales was due to other factors.

This is the line which estimates X from a known value of Y. To obtain its equation we subtitute X for Y, and Y for X in the two previous formulae F. 11.2 and F. 11.3. That is:

Regression of X upon Y

$$\Sigma X = na + b\Sigma Y$$
$$\Sigma XY = a\Sigma Y + b\Sigma Y^2.$$

Substituting the values in our table, we now have:

$$160 = 5a + 285b \qquad \text{Eq.(11.5)}$$
$$9340 = 285a + 16\ 575b \qquad \text{Eq.(11.6)}$$

Proceeding as before:

1. Multiply Eq. 11.5 by 57 (to make the a values equal)

$$9120 = 285a + 16\ 2456 \qquad \text{Eq.(11.7)}$$

2. Subtract Eq. 11.7 from Eq. 11.6

$$220 = 330b$$

$$\frac{220}{330} = b \quad \text{i.e. } b = \tfrac{2}{3}$$

3. Substitute this value in Eq. 11.5

$$160 = 5a + 190$$

$$-30 = 5a$$

$$\frac{-30}{5} = a \quad \text{i.e. } a = -6$$

The equation to the regression line is:

$$X = -6 + \tfrac{2}{3}Y \qquad \text{Eq.(11.8)}$$

As before, we can use this to estimate X, and compare it with the actual data. Taking the day when $Y = 45$ (£ hundreds) we have:

$$X = -6 + (\tfrac{2}{3} \times 45) = -6 + 30 = 24°$$

This agrees exactly. In other words, the plot for this day will lie on the regression line.

Figure 11.3 shows the actual plots, together with the two regression lines.

Note 1. The two lines intersect at a point represented by the means of the two series:

$$(Y = 57; X = 32)$$

Note 2. Although the two scales have been started from zero (to show the points of intersection with the axes), this is not necessary, unless the question calls for it. If we had started the X scale at 16, and the Y at, say, 40, we could have used a more open scale and made the difference between the two lines more pronounced.

Correlation coefficients   Although regression analysis provides a useful method for estimating one variable from another, we still lack a means for giving a numerical value to the correlation present between two sets of data. This is particularly necessary when we wish to make comparisons.

One such measure is the *Pearsonian coefficient of correlation* (sometimes called the product–moment coefficient) represented by r. This is an extension of the method employed for standard deviation.

Application to ungrouped data   There are a variety of formulae for calculating r, each representing a different approach. The one we shall use for the moment is:

$$r = \frac{\Sigma xy}{\sqrt{\Sigma x^2 . \Sigma y^2}} \qquad \text{(F. 11.4)}$$

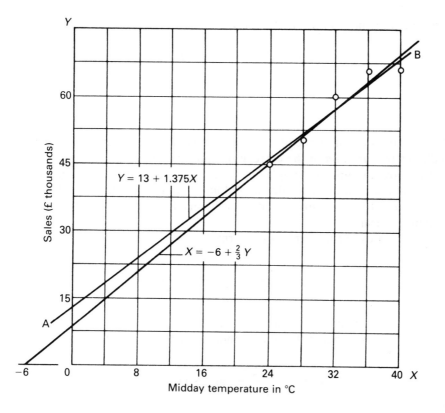

Fig. 11.3 Regression lines for the data in Table 11.1

where x = the *deviation* of each X value from the arithmetic mean of the Xs, and y = the deviation of each Y value from the arithmetic mean of the Ys. (Σ = 'the sum of', as before).

Example 3:

Table 11.2 Output and machines employed in a factory

1	2	3	4	5	6	7	8	
Week number	Number of machines working X	Output (thousands) Y	Deviation from A.M. = 8 x	Deviation from A.M. = 6 y	x^2	y^2	Column 4 × Column 5	
							−	+
1	7	5	−1	−1	1	1	—	1
2	10	8	+2	+2	4	4	—	4
3	6	3	−2	−3	4	9	—	6
4	8	6	0	0	0	0	—	—
5	9	8	+1	+2	1	4	—	2
TOTALS	40	30	—	—	10	18	—	13
A.M.	8	6					$\Sigma xy = +13$	

Substituting in our formula (F. 11.4) we have:

$$\Sigma xy = 13 \quad \Sigma x^2 = 10 \quad \Sigma y^2 = 18$$

Hence $r = \dfrac{+13}{\sqrt{10 \times 18}} = \dfrac{+13}{\sqrt{180}} = \dfrac{+13}{13.42} = +0.9687$

Note. The sign of the end column should always be inserted, since this indicates the *direction* of the correlation (positive or negative).

The above example is a simple one, since in each case the arithmetic mean is a whole number. In practice, this is seldom so, but a modification of the short method for the standard deviation will enable us to work from assumed arithmetic means.

In such cases, we use a modified formula:

$$r = \frac{\Sigma xy}{n \times S.D._x \times S.D._y} \qquad \text{(F. 11.5)}$$

where n = the number of pairs of items.

It will be remembered that the method of calculating S.D. is:

$$S.D._x = \sqrt{\frac{\Sigma x^2}{n}} \quad \text{and} \quad S.D._y = \sqrt{\frac{\Sigma y^2}{n}}$$

where x and y = the deviations from the *true* arithmetic mean in each case.

Substituting these in (F. 11.5) gives us:

$$r = \frac{\Sigma xy}{n \sqrt{\dfrac{\Sigma x^2}{n}} \times \sqrt{\dfrac{\Sigma y^2}{n}}} = \frac{\Sigma xy}{n \dfrac{\sqrt{\Sigma x^2 \times \Sigma y^2}}{n}}$$

Cancelling out the ns gives us the previous formula (F. 11.4).

Example 4 (Table 11.3):

1. In columns 3 and 4 we are taking the deviations from an *assumed* arithmetic mean, since in each case the *real* arithmetic mean is not a whole number. This is the short method, as previously applied to the standard deviation (see Chapter 10).

   Note. A useful tip is to take, as the assumed arithmetic mean, the whole number *immediately below* the real average (in this case, 7 and 14 respectively). This has the advantage of making the subsequent correction always positive, thus avoiding confusion regarding signs at a later stage.

Table 11.3

1	2	3	4	5	6	7	
X	Y	Deviation from assumed A.M. = 7 (= x)	Deviation from assumed A.M. = 14 (= y)	x^2	y^2	$(x) \times (y)$ Column 3 × Column 4	
						−	+
10	9	+3	−5	9	25	15	—
3	20	−4	+6	16	36	24	—
8	15	+1	+1	1	1	—	1
12	6	+5	−8	25	64	40	—
4	22	−3	+8	9	64	24	—
6	18	−1	+4	1	16	4	—
5	23	−2	+9	4	81	18	—
9	12	+2	−2	4	4	4	—
8	16	+1	+2	1	4	—	2
11	8	+4	−6	16	36	24	—
76	149	+6	+9	86	331	153	3
A.M. = 7.6	14.9						3
							−150

2. To get the standard deviation of the Xs, we have:

$$\text{S.D.}_x = \sqrt{\frac{86}{10} - \left(\frac{+6}{10}\right)^2} = \sqrt{8.6 - 0.36} = \sqrt{8.24}$$

There is no need to work this out, because we are not interested in the answer for its own sake.

3. Similarly, we have the S.D. of the Ys:

$$\text{S.D.}_y = \sqrt{\frac{331}{10} - \left(\frac{+9}{10}\right)^2} = \sqrt{33.1 - 0.81} = \sqrt{32.29}$$

4. The student should note most carefully that the totals of columns 3 and 4 give us the real arithmetic mean of the Xs and Ys. They therefore provide a check upon the accuracy of our deviations, which should always be carried out before proceeding, i.e.:

$$\text{A.M. of the Xs is } 7 + \frac{6}{10} = 7.6 \text{ (column 1)}$$

$$\text{A.M. of the Ys is } 14 + \frac{9}{10} = 14.9 \text{ (column 2)}$$

5. We have now corrected out two standard deviations for the fact that we did not work from the real averages, but the total of column 7 is also wrong, for the same reason. The correction for this will be more easily understood if we have another look at formula F. 11.5:

$$r = \frac{\Sigma xy}{n \times \text{S.D.}_x \times \text{S.D.}_y}$$

Since we already have the two standard deviations, this leaves us with $\dfrac{\Sigma xy}{n}$

which is called the *co-variance*. This is the bit which needs correcting, and we do this by subtracting the product of the two correction factors already used in the standard deviations: i.e. 6/10 and 9/10.

Hence we have:

$$\frac{-150}{10} - \left(\frac{+6}{10} \times \frac{+9}{10} \right) = -15 - 0.54 = -15.54$$

Note. In this example, the total of column 7 is *also negative*, so the two numbers must be added. If column 7 had been positive, the correction factor would be taken away, and the final answer would be less.

6. We can now restore the two standard deviations in the original formula, giving us:

$$r = \frac{-15.54}{\sqrt{8.24} \times \sqrt{32.29}} = -0.9526$$

The student may use logarithms (or a calculator) to work this out, and may wonder how the log of a minus quantity can be found. The negative sign shows only the *direction* of the correlation, and may be ignored for purposes of working. It must, of course, be restored in the final answer.

Calculation for grouped data

Example 5 (Table 11.4):

Note 1. Where there are open-ended groups, we first close them. In general, assume that they are the same size as the rest, unless there is evidence to the contrary.

Note 2. As with all grouped distributions, we must assume that the frequencies are concentrated at the midpoints of each group.

Note 3. We use short methods, i.e. work from an assumed arithmetic mean (A.A.M.) (one of the midpoints), and state the deviations in units equal to the group interval, for each variable.

Table 11.4 Days lost by workers according to their ages

(Y)	(X) Number of workers per age group					
Days lost	20–30 yrs	30–40 yrs	40–50 yrs	50–60 yrs	60–	Totals
1–2	4	—	4	—	—	8
3–4	8	8	4	4	—	24
5–6	4	12	16	—	—	32
7–8	—	4	8	4	—	16
9–10	—	—	12	—	—	12
11–	—	—	—	7	1	8
TOTALS	16	24	44	15	1	100

Step 1 Find the standard deviation of the X series.

Table 11.5

1	2	3	4	5		6
Age groups (yrs)	Midpoints	Deviation from A.A.M. of 45 yrs (units of 10 yrs)	Fre-quency	Column 4 × Column 3 Freq. × dev.		Column 5 × Column 3 Freq. × dev.^{2}
				−	+	
20–30	25	−2	16	32	—	64
30–40	35	−1	24	24	—	24
40–50	45	0	44	—	—	—
50–60	55	1	15	—	15	15
60–70	65	2	1	—	2	4
TOTALS	—	—	100	56	17	107
				−39		

$$\text{S.D.}_{x} = \sqrt{\frac{107}{100} - \left(\frac{-39}{100}\right)^2} = \sqrt{1.07 - 0.1521} = \sqrt{0.9179}$$

Note. The answer is in class units of 10 years, and we leave it in this form, since we are not interested in the answer, as such.

Step 2 Find the standard deviation of the Y series.

165

Table 11.6

1	2	3	4	5		6
Days lost	Midpoints	Deviation from A.A.M. of $5\frac{1}{2}$ (units of 2 days)	Fre-quency	Column 4 × Column 3 Freq. × dev.		Column 5 × Column 3 Freq. × dev.^{2}
				−	+	
1–2	$1\frac{1}{2}$	−2	8	16	—	32
3–4	$3\frac{1}{2}$	−1	24	24	—	24
5–6	$5\frac{1}{2}$	0	32	—	—	—
7–8	$7\frac{1}{2}$	1	16	—	16	16
9–10	$9\frac{1}{2}$	2	12	—	24	48
11–12	$11\frac{1}{2}$	3	8	—	24	72
TOTALS	—	—	100	40	64	192
					+24	

$$\text{S.D.}_{\cdot y} = \sqrt{\frac{192}{100} - \left(\frac{+24}{100}\right)^2} = \sqrt{1.92 - 0.0576} = \sqrt{1.8624}$$

Again we leave the answer in class units.

Step 3 We must now find the co-variance, as in the previous example. In the case of a table, which is really a combination of two separate frequency distributions, the method is to multiply together the respective deviations of each distribution, and then to multiply the product by the approximate frequency, as shown in the original data (Table 11.4).

Table 11.7

(Y) Deviations copied from col. 3 of Table 11.6	(X) Deviations copied from col. 3 of Table 11.5				
	−2	−1	0	1	2
−2	4	—	4	—	—
−1	8	8	4	4	—
0	4	12	16	—	—
1	—	4	8	4	—
2	—	—	12	—	—
3	—	—	—	7	1

This is normally done in one operation, using a single table to collect the results, but in order to make clear the steps involved, we shall carry out the operation in two stages.

The figures entered in bold in Table 11.7 are the frequencies copied from Table 11.4.

In the next table (Table 11.8), we multiply the two deviations, and multiply the answer by the appropriate frequency. For example, starting in the top left-hand corner, we have $-2 \times -2 \times 4 = +16$, which we enter as shown below:

Table 11.8

					TOTALS	
					−	+
+16	—	—	—	—		16
+16	+8	—	−4	—		20
—	—	—	—	—	—	—
—	−4	—	+4	—	—	—
—	—	—	—	—	—	—
—	—	—	+21	+6		27
					—	63
					+63	

Note 1. Particular care must be taken over the *sign* of the product, because some deviations are plus and some are minus.

Note 2. In the end columns, we total the figures horizontally, allowing for differences in sign, and enter the grand total at the foot.

According to this result, the co-variance, i.e. $\left(\dfrac{\Sigma xy}{n}\right) = \dfrac{+63}{100}$

but this figure needs correcting, since we have measured the deviations from assumed arithmetic means, instead of using the true ones.

The method of correction is exactly as in the previous example:

$$\text{Corrected co-variance} = \frac{+63}{100} - \left(\frac{-39}{100} \times \frac{+24}{100}\right)$$

i.e. we subtract the product of the two correction factors which we used in the standard deviations—taking care over the signs.

This gives us:

$$+0.63 - (-0.0936) = +0.63 + 0.0936 = +0.7236 \text{ (in class units)}$$

Referring to formula F. 11.5, we have:

$$r = \frac{co\text{-}variance}{\text{S.D.}_x \times \text{S.D.}_y}$$

$$\text{Hence } r = \frac{+0.7236}{\sqrt{0.9179} \times \sqrt{1.8624}}$$

So far, all the calculations have been in units based upon the respective class intervals. A very useful feature of the correlation coefficient is that the answer is independent of the units used—i.e. the units cancel out. Hence, no further correction is needed; the answer is a pure number and is not in any unit at all.

To show the use of logarithms in such cases, we give the working in full:

Log $0.7236 = \bar{1}.8595$

$$\text{Log } 0.9179 = \bar{1}.9628$$
$$\text{Log } 1.8624 = 0.2700$$

Add   0.2328

Square root ($\div 2$)   0.1164

Subtract this from the numerator: $\bar{1}.8595$
   0.1164

$\bar{1}.7431$

Anti-log $\bar{1}.7431 = +0.5535$

Note. The two logs are added *before* being divided by two. This saves time, since we are, in effect, finding one square root (of the product) instead of calculating each individually. The significance and meaning of this result will be discussed under 'Use and interpretation'.

Rank coefficient of correlation (Spearman coefficient)

It often happens that actual numerical values for the two variables, X and Y are not available, or can only be obtained by the expenditure of much time and money.

In such cases, it may be possible to arrange each of the two sets of items in order of merit or importance. This is called *ranking*. For example, if we had 10 trainees and wished to examine the relationship between their command of English and their competence on the job, it would be possible for the supervisor or some similar superior, to write down the names of the 10 persons, and then to allot to each individual a number, which showed his or her position in the ranking. The number given to any particular person is called the *rank or rating*. There would have to be two such rankings—one for English, and the other for competence—and these lists of numbers would correspond to our Xs and Ys in the previous method.

It is, of course, essential to maintain the same order for the *names* of the individuals throughout.

The rank coefficient is calculated from the following formula:

$$R = 1 - \frac{6 \times \Sigma(\textit{differences})^2}{n(n^2 - 1)} \qquad \text{(F. 11.6)}$$

where R = the rank coefficient of correlation; the *differences* are, for each individual item, the numerical difference between the two ranks—ignoring signs; n = the number of items in each; Σ = 'the sum of', as previously.

Example 6:
Nine trainees are given an intelligence test, and are also ranked by their supervisor, according to their ability on the job. The results are given in Table 11.9. Find a measure of correlation between the two results.

Table 11.9 Intelligence ratings for 9 trainees

1	2	3	4	5
Trainee	Rating in test	Rating on job	Difference (Columns 2–3)	Difference²
A	4	5	1	1
B	2	1	1	1
C	6	6	0	0
D	8	7	1	1
E	1	4	3	9
F	9	8	1	1
G	7	9	2	4
H	5	3	2	4
I	3	2	1	1
			TOTAL	22

Applying formula F. 11.6, we have:

$$R = 1 - \frac{6 \times 22}{9(81 - 1)} = 1 - \frac{6 \times 22}{9 \times 80} = 1 - \frac{11}{60} = \frac{49}{60} = 0.817$$

Use and interpretation

Regression lines

We have seen that there are, in general, two such lines. Either one of them can be used to give an estimate of the other variable, and the greater the degree of correlation between X and Y, the closer our estimates will be to the actual values. It can be shown that the angle between the two lines diminishes as the correlation increases, and in the case of perfect correlation, this angle becomes zero—i.e. the two regression lines coincide, and become one. In such cases it does not matter whether we estimate X and Y, or vice versa.

When there is no correlation, the two lines are at right angles to each other (the largest possible angle). One is parallel to the X axis, and the other is parallel to the Y axis.

Thus, it can be seen that there is a close connection between regression lines and correlation, and this can be expressed in mathematical terms (see correlation coefficient).

Interpolation and extrapolation

The great practical use of regression lines is the ability to predict one variable from a known value of the other one. This is valuable in two ways.

First, we may choose some *intermediate* value for our known variable (i.e. something between the two ends of the scale) and use this to get an estimate for the other variable; for example, in Table 11.1, the midday temperatures range between 24 and 40°C. What sales of ice-cream can we expect for, say, a temperature of 30°C? We simply substitute this temperature (our X) in Eq.(11.4) and get an estimate of Y (ice-cream sales). This can be helpful in determining production schedules, stocks carried, etc., using weather forecasts and available statistics of average temperature for the month in question. This is *interpolation*.

Second, we can project our sales, and imagine that the regression line is extended (in either direction). In other words we can choose some *external*, or extreme value for our X, which we have never met in actual practice, and use this as the basis for estimating a value of Y. This is *extrapolation*.

There would not be much point in estimating ice-cream sales for, say, a temperature of 10°C below freezing, or 80°C at the other end of the scale; but in research and scientific work, the ability to make such predictions is valuable. For example, a certain characteristic of a metal or alloy is known to vary with temperature. How will it stand up to extremes of temperatures such as those met with in space flight? The best estimate, short of an actual test at the desired temperature, is to extrapolate a known regression line.

The slope of a regression line, represented by the value of b in its equation (F. 11.1), is sometimes called the *regression coefficient*. In the case of line (A) in Fig. 11.3 on p. 161, this has a value of 1.375 (in £ hundreds). This tells us the change in our Y (in this case ice-cream sales) brought about by a change of one unit in our X (in this case 1°C). That is, for each change (+ or −) of one degree in temperature, we would expect ice-cream sales to change by £137.5.

The other line (B) will, of course, have a different slope, and consequently a different value of b (see Fig. 11.3 on p. 161).

The correlation coefficient

We have said that there is a mathematical relationship between this measure, and the regression lines. In fact, if the slopes of the two lines are represented by b_1 and b_2, then the coefficient of correlation (r) is given by:

$$r = \sqrt{b_1 \times b_2} \qquad \text{(F. 11.7)}$$

With perfect correlation, b_1 and b_2 are the same, because the two lines coincide. Hence:

$$r = \sqrt{b \times b} = b$$

Whenever the student is asked to draw one or more regression lines, in addition to finding their equations, this is a simple matter, if it is remembered that they are always *straight* lines, and that only two points are needed to establish a straight line.

If the equation is of the form $Y = a + bX$, let $X = $ zero. This makes $Y = a$—i.e. it gives the point at which the line cuts the vertical scale.

Choose as the other point a value of X which is some distance away from the first one, depending upon the magnitude of the X series, and find the corresponding value of Y.

Finally, join the two points with a straight line.

It can be proved that the value of r varies between $+1$ and -1. Perfect correlation is represented by unity. (The sign indicates whether it is positive correlation or negative.) A value of 0 indicates no correlation. **Interpretation of r**

Values of unity or zero are very rare in practice, and typical figures are usually of the order of 0.6 to 0.9. What does such a figure mean? The value of r is very much affected by the size of the sample (i.e. the number of items) and for small samples it should be treated with great reserve. Tables are available which show the size of sample needed for a given level of confidence in the result.

As a generalization, one can say that for small samples a value below 0.5 could have arisen by chance; i.e. it is not significant, and could disappear in further samples.

The standard error of r is discussed later, in Chapter 17.

The fact that the calculation of r is independent of the units used is important in two ways.

First, the student need not understand what the Xs and Ys mean. It would not matter if they were expressed in units never heard of before. The calculation of r is purely mechanical.

Second, it is possible to vary the units at will to simplify the calculation. For example, if one series is in decimals, multiply it by 10 (or 100) to remove them. If one series has 0s at the end of each figure, divide by 10 (or 100) to get rid of them. Provided all the numbers in any *one* series are treated alike, they can be multiplied or divided at will—without any further correction being necessary.

Rank correlation (R) is used when actual values for X and Y are not available. **Rank correlation** It is frequently impossible to assign numerical values to a variable. For example, there is no numerical measure of shades of colour, or of character aspects such as good temper, dependability, and so on. It may be possible, however, to rank a group of individuals with regard to some quality, or to arrange various shades of colour in order of vividness. Like the Pearsonian coefficient, the value of R varies between $+1$ and -1, but the results obtained by the two methods are not necessarily comparable.

A difficulty sometimes arises when two or more items are ranked equally; i.e. given the same position number. Strictly speaking, the method of calculation does not provide for this, but when met with in an examination, there are two possible methods of dealing with it.

1. If two individuals are, say, equal sixth rank, count each as six, and the next ranking item as eight (in other words, there is no seven).
2. Alternatively, average the two (or more) items. For example, if ranked equal sixth, count one as sixth and the other as seventh. Their average rank is then:

$$\frac{6 + 7}{2} = \frac{13}{2} = 6\frac{1}{2}$$

The next ranking item is eight as in the previous method.

Needless to say, if both the X and Y series need adjustment in this way, the method used should be the same in each case, preferably method 2.

The ranking method can also be used for evaluating personality tests, etc. Supposing we wish to know whether a particular psychological test is useful for our purpose. We take a group of people, give them the test, and rank them according to their performance. If the test is supposed to measure ability of some kind, we then ask the supervisor, or other person who knows them, to rank the same individuals according to their actual ability on the job. We then compare the two rankings to see whether there is any correlation.

Correlation and regression and computer

MINITAB example   Correlation and regression are easily accomplished with MINITAB. The following example should work on both mainframe and PC versions of MINITAB. Having called up the package with MINITAB ⟨**RETURN**⟩, we have input the data from the keyboard with the READ command. PC MINITAB users may choose to input the data using the 'data editor' mode (accessed with ⟨**Esc**⟩). We have used the ice-cream sales data from earlier in the present chapter. For all MINITAB users it is possible to READ the data into the worksheet from a previously created and stored data file (see Chapters 6 and 7).

```
MTB > READ C1 C2
DATA> 6500 40
DATA> 4500 24
DATA> 6500 36
DATA> 5000 28
DATA> 6000 32
DATA> END

MTB > REGRESS C1 1 C2
```

The REGRESS command asks to regress or *predict* the data in C1 from the *single* predictor (hence the '1' between the column designations), the data in C2. Below, we reproduce the most important parts of the output:

```
The regression equation is
C1 = 1300 + 138 C2
```

Pearson correlation is obtained simply with the CORRELATION command, which may be shortened to CORR. The single line of output is also reproduced below:

```
MTB > CORR C1 C2

Correlation of C1 and C2 = 0.957
```

Spearman's rank order correlation can be calculated by ranking the data in the two columns, and then using the CORR command, e.g.:

```
MTB > RANK C1, C3
MTB > RANK C2, C4
MTB > CORR C3 C4
```

Using the RANK command we have to specify the columns into which we want the ranked versions of the variables to be placed. Here C1 ranks are placed in C3, and C2 in C4. Try this procedure, inspect the data with PRINT C1–C4 (or in data editor mode), and compare the Pearson and Spearman results.

Using SPSS/PC+

SPSS/PC+ is the PC version of SPSS[x]. It is complex not only because SPSS software offers many data manipulation routines but also because SPSS/PC+ gives us three ways of inputting commands from the keyboard:

1. Selection of the command from a menu.
2. Typing the command into a file (compare SPSS[x]).
3. Typing the command at the 'SPSS/PC>' command prompt.

SPSS/PC+ is an interesting mixture of 'interactive' and 'batch' working (see Chapter 3). Method 3. is the most obviously interactive, and is like the input of commands at the 'MTB >' prompt in MINITAB—the package executes the commands and displays results immediately.

It will help SPSS/PC+ users to have first read the SPSS[x] sections in the previous chapter. When we first enter SPSS/PC+ (see below for system commands to do this), we are confronted by three *windows*. The smallest window (top left) is the *menu window*, offering a list or menu of commands that can be highlighted and selected. The menu item currently highlighted by the highlight bar is explained in the *help window* alongside (top right). The bottom half of the screen is the *scratchpad window*, the file into which our commands are placed once typed or selected from menu. Fig. 11.4 shows an SPSS/PC+ screen with commands already entered into the bottom scratchpad.

Selecting commands from menu can be slower than simply typing them. Menu selection is also rather confusing (there are too many choices) for SPSS/PC+ beginners. We therefore recommend typing commands direct from the keyboard. In the example that follows we have chosen to use method 2. because it is simple, and it allows easy correction of typing errors before the commands are executed.

Once we are inside SPSS/PC+ there is a 'help' system that is called by pressing the ⟨F1⟩ key. Then press ⟨**RETURN**⟩ or else use the cursor 'arrow' keys to highlight the topic of interest on the menu bar at the very bottom of the screen. The highlighted item is selected with ⟨**RETURN**⟩. You can exit the help system with ⟨**Esc**⟩. It is easy to get lost. Beginners may find that it is better to avoid the help system!

SPSS/PC+ EXAMPLE

We shall use SPSS/PC+ in an uncomplicated way. Our task is to recalculate the Pearson correlation coefficient for the output and machines data of *Example 3* on page 161. On the hard disk of your PC, the package is probably stored within a *directory* (an area for working and storage) called 'SPSS'. First gain access to this directory. The command is most likely to be:

```
cd spss
```

When the prompt changes, call the SPSS/PC+ package by keying:

```
spsspc
```

Once SPSS/PC+ has loaded successfully you will see the screen divided into the three windows. We shall use method 2. of command entry—keying commands into a file.

Resist all temptation to panic at the sight of so much complex information displayed on the screen at the same time! We shall completely ignore the two menus in the top half of the screen. Instead, we shall type our SPSS/PC+ commands directly into the lower half, the scratchpad. To do this we disable the menu window and enter edit mode by holding down the ⟨**Alt**⟩ key and pressing the '**E**' key at the same time:

```
⟨Alt⟩ + ⟨E⟩
```

We can now type our commands into the scratchpad, pressing ⟨**RETURN**⟩ at the end of each line. Our first SPSS/PC+ 'run' or 'job' begins with DATA LIST and ends with the CORRELATION procedure. The whole screen, including our scratchpad input, is reproduced below in Fig. 11.4.

Notice that the SPSS/PC+ job is similar to the SPSS[x] examples previously encountered. However, there are important differences. No FILE= subcom-

```
      MAIN MENU                      orientation

orientation                The "orientation" section provides a
read or write data         brief explanation of how the SPSS/PC+
modify data or files       Menu and Help system works. If you have
graph data                 not used the Menu and Help system
analyze data               before, you may want to read through
session control &          the screens in the orientation.
info
run DOS or other pgms      To do so, press    (Enter).

extended menus             Part A of the SPSS/PC+ 4.0 manual
SPSS/PC+ options           contains a more complete introduction
FINISH                     to the Menu and Help system.
                           F1=Help    Alt-E=Edit Alt-M=Menus   on/
      (Esc for menu)       off


DATA LIST /
                  MACHINES 1-2, OUTPUT 3-6.
BEGIN DATA.
075000
108000
063000
086000
098000
END DATA.
CORRELATION        MACHINES OUTPUT
                                      Ins          Std Menus
                                scratch.pad
```

Fig. 11.4 Screen from SPSS/PC+ showing machines/output example

mand is used with DATA LIST. There is a major addition, however. The grammar of SPSS/PC+ demands that each command ends with a full stop, and you will find that the run will fail if these are omitted. In our example above there are four commands and four full stops. In other regards SPSS/PC+ grammar is more forgiving than that of SPSS[x].

The DATA LIST command keyphrase is followed by a forward oblique ('/') and then the *specification* of the two variable names ('MACHINES', 'OUTPUT') with their column locations. The data are sandwiched between the BEGIN DATA and END DATA commands. The CORRELATION command is simply followed by the specification of the variables used.

So far, we have only typed the commands into a temporary file. None has yet been acted upon. To activate the run we move the cursor to anywhere on the first line (DATA LIST) and press the 'run from' function key ⟨**F10**⟩. We confirm the 'run from cursor' choice with ⟨**RETURN**⟩. SPSS/PC+ processes the commands and data, and outputs results to the screen. We move from screen to screen of results by pressing any key, e.g. ⟨**RETURN**⟩. Below, we reproduce a slightly reduced form of the output:

```
    Correlations:    MACHINES    OUTPUT

    MACHINES         1.0000        .9690
         OUTPUT                .9690      1.0000

    N of cases:         5
```

What we see is a simple *correlation matrix*. There is really only one correlation of interest here, the coefficient 0.9690. The correlation matrix correlates each variable with every variable on the CORRELATION command specification. The correlation of MACHINES with itself is necessarily perfect (+1.0000), and the same for OUTPUT. The correlation of MACHINES with OUTPUT is the same as that of OUTPUT with MACHINES, hence the double appearance of the 0.9690 figure.

At the end of the output sequence we return to the three-window screen display. We wish now to produce a scattergram of the data. We do this with the PLOT command. But first we must return to the edit mode with ⟨**Alt**⟩ + ⟨**E**⟩. We then key, *without* pressing ⟨**RETURN**⟩:

```
    PLOT /      PLOT OUTPUT WITH MACHINES.
```

The PLOT command is followed by the subcommand PLOT OUTPUT, plus the specification of the variables linked by 'WITH'. Do not forget to end with the full stop! At this point we can run the command line without having to rerun the data entry lines—SPSS/PC+ will have remembered the data from the first run.

The cursor is already on the line of the PLOT command. We therefore simply press ⟨**F10**⟩, the 'run from' function key, and confirm 'run from cursor' with ⟨**RETURN**⟩. The output is reproduced below:

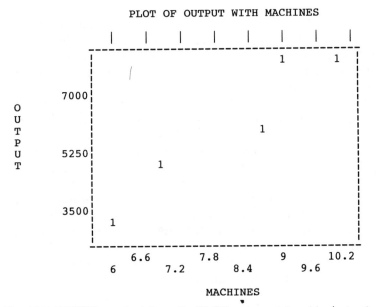

Fig. 11.5 SPSS/PC+ output from the PLOT command (machine/output)

SPSS/PC+ users who wish to rework the examples given for SPSS^x in the previous chapter will find that frequency tables and descriptive statistics (averages, measures of dispersion) are easily produced. The procedure FREQUENCIES displays a frequency table, and descriptive statistics can be added with the STATISTICS subcommand. FREQUENCIES is appropriate to *discrete* data (see Chapter 5). The procedure DESCRIPTIVES replaces the SPSS^x CONDESCRIPTIVES command, and is more suited to *continuous* data, or data where there are many values. On the data in this chapter you might try:

```
FREQUENCIES      MACHINES / STATISTICS.
   DESCRIPTIVES     / VARIABLES OUTPUT.
```

Note that we have followed FREQUENCIES with the STATISTICS subcommand. DESCRIPTIVES *requires* the VARIABLES subcommand. To exit from SPSS/PC+ we need the FINISH command. Once the three-window screen reappears we re-enter the edit mode with ⟨**Alt**⟩+⟨**E**⟩. We key:

```
FINISH.
```

With the cursor on the FINISH command line, we press the ⟨**F10**⟩ key and confirm with ⟨**RETURN**⟩. We exit the SPSS/PC+ package and return to the operating system of the PC.

Users of SPSS/PC+ who need to manipulate or 'transform' data (COMPUTE, IF, RECODE) or to select data (SELECT IF) will find these topics covered in the final sections of Chapter 15.

SPSS^x users will find that they can reproduce the correlation/scattergram example above with an SPSS^x command file very similar to the SPSS/PC+ example. The SPSS^x commands (without repetition of data) are: **SPSS^x version**

```
DATA LIST      FILE = INLINE
               / MACHINES 1-2, OUTPUT 3-6
   BEGIN DATA
     .
     .
   END DATA
   CORRELATION    MACHINES OUTPUT
   PLOT           PLOT OUTPUT WITH MACHINES
   FINISH
```

Note that SPSS^x commands do not take a final full stop. The PLOT procedure also lacks the '/' between command and subcommand. In recent versions of SPSS^x, the SCATTERGRAM procedure has been replaced by PLOT. The older procedure of PEARSON CORR may still work on versions of SPSS^x that also support CORRELATION.

12

Time series

Method

A time series is the name given to a series of figures recorded through time. The name given to the graph on which such a series is plotted is a *historigram* (i.e. a history or record over time).

This should not be confused with the histogram, dealt with in Chapter 7.

The series may be plotted at daily, weekly, monthly, yearly or any other intervals of time, and the horizontal axis is always chosen as the time axis.

If we were to plot the graph of a time series consisting of monthly or quarterly figures, over a large number of years, we might get something like Fig. 12.1.

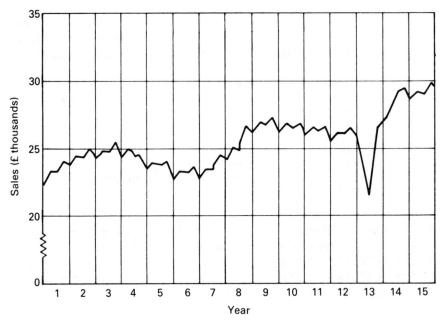

Fig. 12.1 Quarterly sales (£ thousands) of a firm

This actually represents a mixture of various influences, the principal ones being:

1. *The long-term trend* It can be seen that the overall picture is one of expansion, since the sales have increased from some £22 000 per quarter at the start, to a figure of roughly £30 000, at the end of the period.
2. *The cyclical movement* This is the wave-like formation, generally due to the influence of booms and slumps on business activity. In the example, the period of each complete cycle, i.e. the distance in time from one peak to the next, is roughly 6 years but, in practice, this period may be anything from 5 to 14 years.
3. *The seasonal variations* These are the small steps in each year, and are due to the fact that the quarterly figures follow a seasonal pattern, i.e. certain quarters are usually busy ones, while others show less active sales. This pattern will obviously vary from one industry to another.
4. *The non-recurring influences* These may be good or bad, but since they do not occur with any statistical regularity, they cannot be measured or predicted.

    Such an influence may be seen in Year 13, where there is an unusual dip in the graph. This may have been the time when the firm's factory was destroyed by fire! Or it may have been some political event, such as a threat of war, or a financial crisis.

It is essential that we should be able to disentangle these various influences, and to measure each one separately. This procedure is known as the *analysis of a time series*.

Example 1:
The figures given in the table below are a record of withdrawals over a period of years for a bank in an industrial area:

Table 12.1 Withdrawals from a bank in an industrial area

Year	7	8	9	10	11	12	13	14	15
Withdrawals (£ten thousands)	25.8	19.2	26.0	27.1	22.8	28.7	28.5	24.5	29.6

A graph of this series may be drawn as shown in Fig. 12.2.

It can be clearly seen that the curve of the series varies with occasional troughs and peaks every few years. There is, in fact, what may be called a cycle (trough to trough, or peak to peak). This cyclical movement, however, is not level for the whole of the period. It seems to be superimposed on a steady upward trend. The word 'trend' means a long-term movement, irrespective of occasional variations in the short period.

179

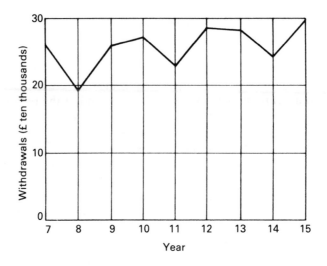

Fig. 12.2 Withdrawals from a bank in an industrial area

If we could somehow split these movements—the *cycle* and the *trend*—the information might be of use to the bank manager. The trend might suggest whether business is expanding or contracting in the long run. If the trend is upwards (as in our example), more staff and counter space, or more investment in office machinery, may be needed. The cycle may suggest the possible interval of years during which cash is being taken out of, or left to accumulate in, the bank. This may lead the bank manager to adjust the bank's policy in giving loans for two-, three-, or five-year periods.

The trend

We can separate the trend by a statistical technique known as the *moving average*. Table 12.2 below shows the method.

In the table are columns for the years (column 1), and for the withdrawals originally given (column 3). To use the moving averages method, we take successive averages of the original figures over a fixed number of years. In this cast it has been decided to take three years as the fixed number. (How to make this decision is explained later—see 'Use and interpretation' on p. 188). Column 2 simply shows the years re-labelled in threes (a, b, c).

Column 4 shows the totals of each set of three successive years, e.g. Year 7, Year 8 and Year 9 add up to 71.0, and this figure is placed opposite the centre of these three years, i.e, opposite Year 8 in column 4. We proceed by adding the next set of three years, i.e. dropping the first year (Year 7) and adding the next (Year 10). This total is 72.3, and it is placed opposite Year 9, and so on.

Column 4 is completed when we have total figures opposite all the years except the first (Year 7) and the last (Year 15), because moving averages could not be calculated for those years.

From the totals of threes (column 4) we can now work out the moving average, or trend, in column 5, by dividing each total by three, e.g. Year 8:

Table 12.2

		Trend			Variation
Column 1	Column 2	Column 3	Column 4	Column 5	Column 6
Year	Period of moving average	Withdrawals (£ten thousand)	Totals of three	Moving average of three	Variation from the trend
7	a	25.8	—	—	—
8	b	19.2	71.0	23.7	−4.5
9	c	26.0	72.3	24.1	+1.9
10	a	27.1	75.9	25.3	+1.8
11	b	22.8	78.6	26.2	−3.4
12	c	28.7	80.0	26.7	+2.0
13	a	28.5	81.7	27.2	+1.3
14	b	24.5	82.6	27.5	−3.0
15	c	29.6	—	—	—

$71.0 \div 3 = 23.7$ (rounded to first decimal place). The name 'moving average' is a literal description of the method just used, where the *average* of a fixed block of three years is taken and *moved* over the entire period.

The moving average, or trend, is shown in the following graph, together with the graph of the original figures, given previously:

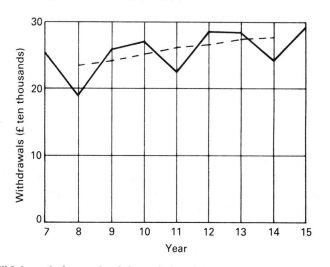

Fig. 12.3 Withdrawals from a bank in an industrial area showing three-yearly moving average

Note. One disadvantage of the moving averages method which can be appreciated at this stage is that, because it is short at both ends, it does not cover the whole period.

The cyclical variation

To work out the second part of the problem, i.e. the cycle, we proceed as follows.

Column 6 is added to the previous table and headed 'Variation from the trend'. This column is calculated by subtracting each trend figure (column 5) from the corresponding figure of actual withdrawals (column 3).

Column 3 minus column 5 = column 6:

$$\text{Year 8:} \quad 19.2 - 23.7 = -4.5$$
$$\text{Year 9:} \quad 26.0 - 24.1 = +1.9$$

and so on.

Special attention must always be paid to the plus and minus signs, but we cannot go wrong if we remember that the variation figure is the variation *from* the trend. Ask yourself, 'Does the actual figure differ from the trend figure positively or negatively?' Actual figures which are *larger* than trend figures show a plus variation sign, and vice versa.

When column 6 has been completed (once again we can have no figures for Year 7 and Year 15), a further step is necessary before we can finally calculate the cyclical variation.

Transfer the variation figures in column 6 to a new table, using the a, b and c labels of column 2 as your headings, thus:

Table 12.3

	Years a	Years b	Years c
	—	−4.5	+1.9
	+1.8	−3.4	+2.0
	+1.3	−3.0	—
TOTALS	+3.1	−10.9	+3.9
Cyclical variation	+1.6	−3.6	+2.0
Adjustment	—	—	—

The yearly variations are totalled in the 'Totals' row, taking careful note of the addition of signs. The next row 'Cyclical variation' is obtained by dividing the totals by the number of items added together in each column. For example:

$$\text{Years a} \quad \frac{+3.1}{2} = +1.6$$

$$\text{Years b} \quad \frac{-10.9}{3} = -3.6$$

$$\text{Years c} \quad \frac{+3.9}{2} = +2.0$$

These are our cyclical variations and they should add up to zero. It can be seen that they do, so there is no need to complete the last row 'Adjustment'. Sometimes, due to the nature of the moving average method, the plus total of variations does not exactly equal the minus total of variations. The adjustment is then calculated as follows:

- If the excess is *positive*, divide it by the number of columns used and *subtract* it from each final variation;
- If the excess is *negative*, divide it by the number of columns used and *add* it to each final variation.

If this is done, the final variations will add up to zero.

Example 2:

The figures given below are a record of the withdrawals from the same bank as in Example 1. This time, however, instead of covering each year from Year 7 to Year 15, the figures cover each quarter from the first quarter (January–March) in Year 13, to the second quarter (April–June) of Year 16. The figures for the last two quarters of Year 16 were not available (N.A.).

Seasonal variation

Table 12.4

Year	Quarters			
	1	*2*	*3*	*4*
13	5.2	6.8	9.1	7.4
14	4.1	5.7	8.2	6.5
15	5.3	6.9	9.4	8.0
16	5.5	7.1	N.A.	N.A.

The graph of the series can be drawn as shown in Fig. 12.4.

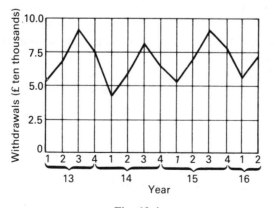

Fig. 12.4

A rising and falling pattern shows itself again in this curve, but this is no longer a cyclical variation (variation within a *number of years*). Our graph shows a *seasonal variation*, or the variation within *separate years*—'season' referring, of course, to the seasons of a year.

As the cyclical variation gave useful information, so this seasonal variation may give the bank manager an idea of the withdrawals to be expected during the various quarters, and thus the amount of cash it is necessary to keep in his till.

We proceed, as before, to separate the two associated movements in the figures. These are the trend and the seasonal variation.

Table 12.5

		Trend				*Variation*
Column 1	*Column 2*	*Column 3*	*Column 4*	*Column 5*	*Column 6*	*Column 7*
Year	Quarter	Withdrawals	Totals four quarters	Centred totals	Moving average ($\div 8$)	Variation from the trend
13	1	5.2	—	—	—	—
	2	6.8	28.5	—	—	—
	3	9.1	27.4	55.9	7.0	+2.1
	4	7.4	26.3	53.7	6.7	+0.7
14	1	4.1	25.4	51.7	6.5	−2.4
	2	5.7	24.5	49.9	6.2	−0.5
	3	8.2	25.7	50.2	6.3	+1.9
	4	6.5	26.9	52.6	6.6	−0.1
15	1	5.3	28.1	55.0	6.9	−1.6
	2	6.9	29.6	57.7	7.2	−0.3
	3	9.4	29.8	59.4	7.4	+2.0
	4	8.0	30.0	59.8	7.5	+0.5
16	1	5.5	—	—	—	—
	2	7.1	—	—	—	—

The first three columns simply show the years, quarters and the original figures of withdrawals. As there are four quarters (or seasons) in the year, we use a moving average of four, therefore column 4 shows the totals of blocks of four successive quarters.

Column 5 shows centred totals, i.e. successive pairs of the totals in column 4 are added, and the results placed opposite the corresponding quarters; thus:

$$28.5 + 27.4 = 55.9$$

This centring is necessary whenever there is an *even* number of items in the moving average—e.g. seasonal variations (four), or even number cyclical variations, in order to ensure that the moving average refers to the original figures of definite quarters or years.

Note. This was not necessary in Example 1, because we took an odd number (three) of years and therefore we had a middle withdrawal figure against which to place each moving average.

In column 6, the centred totals of column 5 are each divided by eight (totals of four × totals of two), and the moving average or trend is obtained.

It should be noted that the first two quarters of Year 13 and the first two quarters of Year 16 do not yield any figures. As before, the trend is shortened at each end, and it should be noted that the greater the number of items in the moving average, the more trend figures are lost at each end of the calculation.

The series can be graphed again and the trend included:

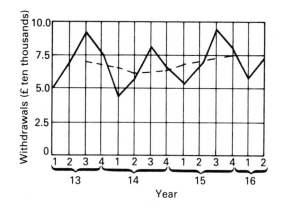

Fig. 12.5 Withdrawals from a bank showing four-quarterly moving average

The seasonal variation is calculated by working out column 7 in the previous table. The trend (column 6) is subtracted from the actual withdrawals (column 3), and the variation (plus or minus) recorded.

The variation figures are subsequently arranged (as in the case of the cyclical variation) in a table:

Table 12.6

Year	Quarters			
	1	*2*	*3*	*4*
13	—	—	+2.1	+0.7
14	−2.4	−0.5	+1.9	−0.1
15	−1.6	−0.3	+2.0	+0.5
16	—	—	—	—
TOTALS	−4.0	−0.8	+6.0	+1.1
Seasonal variation	−2.0	−0.4	+2.0	+0.4

It can be seen that the totals of the final figures, positive and negative, add up exactly to zero, so no adjustment is needed. If this total had differed significantly from zero, then an adjustment would have been made, as described in *Example 1*.

Note. When the calculations are correct only to one decimal place (as in the examples given) it is wrong to introduce any adjustment which would involve additional decimals. The final figures cannot be any more correct than the original ones. In such cases a partial correction may be used. For example, if the total of the seasonal figures were +3.0, dividing this among the four quarters would mean a correction of $\dfrac{3.0}{4} = 0.75$ to be subtracted from each quarterly variation. In this case, a total correction of 2.8 could be applied. This would mean a deduction of $\dfrac{2.8}{4} = 0.7$ from each quarter. The totals would now add to +0.2 which is much better than before.

Least squares method

So far, in this chapter, we have described the analysis of a time series by the method of moving averages, i.e. repeatedly calculating a series of different average values as we moved along an original series, as it were, laying down a carpet of progressive snapshots. We produced a trend (as in Fig. 12.3) which was *not* a straight line. If the original figures had had a cycle which repeated itself *exactly* we should have got a straight line!

In fact it is easy to produce a straight line trend if we prefer to look at the average *overall* rate of increase (or decrease). We can do this by the method of least squares.

In Chapter 6 a line of 'best fit' was drawn on a scatter diagram (e.g. see Fig. 6.9 in Chapter 6). In Chapter 11 another method, the 'three-point method' of applying a straight regression line to two sets of data, was given. Later, in Chapter 11, we gave the formula (F.11.1) for the equation of a straight line:

$$Y = a + bX$$

(see Fig. 11.1) whereby we could *minimize* all positive and negative deviations of the data from a straight line (which could be drawn through the data) by squaring the deviations. This is our method of least squares.

In a time series, one of the two sets of data is, of course, time itself, which increases by equal amounts (years, quarters, etc.). Therefore we can work out a straight line trend by using the above equation if we have time = X, and our data = Y, and squaring only those deviations of X (the irregular data).

Take *Example 1*. 'Withdrawals from a bank in an industrial area' (Table 12.1). We use the two formulae F.11.2 and F.11.3, which are virtually repeated below:

$$\Sigma Y = na + b\Sigma X$$
$$\Sigma XY = a\Sigma X + b\Sigma X^2$$

and we now calculate the necessary figures for *Example 1*:

Table 12.7

Year X	Withdrawals Y	X	XY	X^2
7	25.8	0	0	0
8	19.2	1	19.2	1
9	26.0	2	52.0	4
10	27.1	3	81.3	9
11	22.8	4	91.2	16
12	28.7	5	143.5	25
13	28.5	6	171.0	36
14	24.5	7	171.5	49
15	29.6	8	236.8	64
(n = 9)	232.2 (ΣY)	36 (ΣX)	966.5 (ΣXY)	204 (ΣX^2)

Note. The 'value' of the nine years themselves (Year 7, Year 8, etc.) do not matter—only the increase of the next on the last; hence, in column 3 we write 0, 1, 2, etc. (intervals of one year).

Substituting, in our formulae, for ΣY, ΣZ, ΣXY and ΣX^2, we have:

$$232.2 = 9a + 36b \qquad \text{Eq. (12.1)}$$

$$966.5 = 36a + 204b \qquad \text{Eq. (12.2)}$$

We solve these two equations as follows:

1. Multiply Eq. (12.1) by four (to make the *a*'s equal to those in Eq. [12.2])

$$928.8 = 36a + 144b \qquad \text{Eq. (12.3)}$$

2. Subtract Eq. (12.3) from Eq. (12.2) (to eliminate *a*)

$$37.7 = 60b$$
$$\frac{37.7}{60} = b, \text{ i.e. } b = 0.6 \text{ (to one decimal place)}$$

3. Substitute this value in Eq. (12.1):

$$232.2 = 9a + 21.6$$
$$232.2 - 21.6 = 9a$$
$$210.6 = 9a, \text{ i.e. } a = 23.4$$

The regression line (trend) of least squares is therefore:

$$Y = 23.4 + 0.6X$$

Let us now see how the trend figures obtained from this formula compare with our original series and with the trend figures (Table 12.2) we obtained from the method of moving averages:

Table 12.8

Year	Withdrawals	Moving average of three	Least squares method
Column 1	Column 2	Column 3	Column 4
7	25.8	—	$23.4 + 0.6 \times 0 = 23.4$
8	19.2	23.7	$23.4 + 0.6 \times 1 = 24.0$
9	26.0	24.1	$23.4 + 0.6 \times 2 = 24.6$
10	27.1	25.3	$23.4 + 0.6 \times 3 = 25.2$
11	22.8	26.2	$23.4 + 0.6 \times 4 = 25.8$
12	28.7	26.7	$23.4 + 0.6 \times 5 = 26.4$
13	28.5	27.2	$23.4 + 0.6 \times 6 = 27.0$
14	24.5	27.5	$23.4 + 0.6 \times 7 = 27.6$
15	29.6	—	$23.4 + 0.6 \times 8 = 28.2$

The student will see that the two trend figures (columns 3 and 4) are fairly close substitutes and can guess that the more regular the cycles of the original series, the more closely they would fit with each other.

Use and interpretation

How to choose a moving average

If we are asked to find the seasonal variation and the trend, we are usually given figures for either the four quarters of the year, or figures for the 12 months of the year. It is obvious that, when 'seasonal' means 'within the year', we must use a four-quarterly moving average in the first case, and a 12-monthly moving average in the second case.

Note. Because these are even numbers, centring must be applied.

If the question concerns the cyclical variation, we may choose any number of years. The problem here is, which is the correct number?

The idea behind the separation of the trend is to eliminate the pattern of the shorter season or cycle and to obtain a smooth, long-term movement.

Consider the figures in Table 12.9 to which a five-yearly moving average, and then a seven-yearly moving average have been applied.

Which is the correct one? Quite obviously the seven-yearly moving average. By using a five-yearly moving average, the trend (column 4) still appears to be rather irregular, whereas in the seven-yearly moving average (column 6), the cyclical variation has been removed completely, and we are left with a smooth trend.

The correct method is to determine how many years lie between the troughs or the peaks (i.e. what is the cycle?) and, if the pattern is regular enough, to take this as your moving average. In the example above, it can be seen that the original figures (column 2) are repeated exactly every seven years. This could be seen even more clearly if we were to draw a graph of these original figures.

In Example 1 given previously, the three-yearly moving average was obviously the best one to use, because this was the most appropriate cycle.

Table 12.9

Column 1	Column 2	Column 3	Column 4	Column 5	Column 6
Year	Original data	Five-yearly moving annual total	Five-yearly moving average	Seven-yearly moving annual total	Seven-yearly moving average
1	22	—	—	—	—
2	22	—	—	—	—
3	20	101	20.2	—	—
4	19	97	19.4	140	20
5	18	96	19.2	140	20
6	18	98	19.6	140	20
7	21	101	20.2	140	20
8	22	103	20.6	140	20
9	22	104	20.8	140	20
10	20	101	20.2	140	20
11	19	97	19.4	140	20
12	18	96	19.2	—	—
13	18	—	—	—	—
14	—	—	—	—	—

Limitations of the moving average method

As can be seen from the graphs and tables used previously in this chapter, the trend covers a shorter period than the original figures, and the greater the moving average period, the shorter will be the trend; our calculations will, therefore, be of correspondingly less use.

If calculations are to be really valuable, we should take a long series of figures when calculating the moving average (our examples were necessarily short to save elaborate calculations) and only then, if the pattern is *regular* enough, can we be sure that there are definite, *repeating* factors causing the variations.

Why do we analyse time series?

As explained earlier in the chapter, by separating the seasonal variation, cyclical variation and the trend, we hope to learn something of the behaviour of the series with which we are dealing and probably, also, to use this knowledge as a basis for future action.

A time series can usually be split up into several associated movements. To each type of movement we can assign probable causes and possible future courses of action. The example of our bank can be used to illustrate this:

Table 12.10

Movement	Probable causes	Possible future action
Trend (upwards)	Inflation: rise in volume of money, therefore rise (long-term) in withdrawals. Expansion of business.	Heavier capital investment in buildings, staff, and machines.
Cyclical variation	Trade cycle—rise in withdrawals when money is in short supply and interest rates are low. Unemployment in the area.	Adjust loan policy, e.g. raise interest rates, offer services, give longer loans.
Seasonal variation	Withdrawals high just before Christmas and summer holidays.	Engage extra staff. Increase till-money.

In addition to the trend, cyclical variation and seasonal variation, we may also meet *special variations*. These are due to unusual happenings such as earthquakes, fire, strikes, etc.

It should be noted that, in Example 1, the cyclical variation was separated from a trend, and then in Example 2, from a *shorter series*, the seasonal variation was separated from a trend. The second trend (in Example 2) was a shorter-term trend than that in Example 1, and it included the cyclical variation, as can be seen from the way that this shorter-term trend dips in Year 14 in the second graph (Example 2, Fig. 12.5).

Other uses of the analysis

Time series analysis is widely used in business and commerce for planning ahead in the expectation that previous performance and rhythm will be repeated to some extent. Special fields in which it is used are budgetary control, stockholding, investment in stocks and shares and, particularly, in market research. In the latter field it can be used to try to assess what size of sales can be achieved with a given advertising budget, and how sales are increased with increases in this budget.

Extrapolation

Most of these uses involve the additional process of *extrapolation*.

For instance, in our Example 1, can we tell what will be the probable amount of withdrawals in Year 16?

In Table 12.2, column 5 shows the trend which rises from 23.7 in Year 8 to 27.5 in Year 14. This is a rise of 3.8 over six years, or an average of $\dfrac{3.8}{6} = 0.6$ per year.

If we assume that the trend will continue to rise, the trend figure for Year 16 would be 28.7, i.e.:

Year 14 = 27.5
Year 15 = 27.5 + 0.6 = 28.1
Year 16 = 27.5 + 0.6 + 0.6 = 28.7

We know also the cyclical variation for each of the three years in any cycle (see Table 12.3), and that Year 16 would fall in 'years a' if we continued the table. As the cyclical variation for 'years a' is + 1.6, we simply add 1.6 to the estimated trend figure for Year 16, thus:

Estimated trend + cyclical variation = estimated withdrawal

(28.7)   +   (+1.6)   =   30.3

Therefore the estimated withdrawal figure for Year 16 is £303 000.

Such estimation by extrapolation, although widely practised, is not recommended. Although many economic and commercial series show a fairly regular pattern, it is often a fallacy to suppose that cycles will continue as they have in the past. There are at least two good reasons for this. First, the well-known unpredictability of scientific advance and invention, especially in the twentieth century, and secondly, the fact that the field of government control nowadays is so widespread and deep that government interference (e.g. passing legislation, altering taxation, etc.) may suddenly upset any predictions that natural forces will continue as in the past.

Finally, we must mention a popular way of presenting a time series in a graph; **Seasonal** i.e. that of presenting the 'seasonally adjusted data'. This follows very simply on **adjustment** what we have already learnt. To adjust a series of original figures, we calculate the seasonal variations and simply deduct these from the original figures. Thus, the seasonally adjusted figures of withdrawals in Example 2 (Table 12.5, column 3) would appear (Year 13 to Year 14) thus:

Table 12.11

Column 1	Column 2	Column 3	Column 4	Column 5
Year	Quarter	Withdrawals	Seasonal variation	Seasonally adjusted
13	1	5.2	−2.0	7.2
	2	6.8	−0.4	7.2
	3	9.1	+2.0	7.1
	4	7.4	+0.4	7.0
14	1	4.1	−2.0	6.1
	2	5.7	−0.4	6.1
	3	8.2	+2.0	6.2
	4	6.5	+0.4	6.1

The seasonally adjusted data may now be graphed in the normal way from column 5 of the table.

Comment on the least squares method

Comparison of the moving average method and the least squares method in Table 12.8 shows that the latter gives a trend which (unlike the moving average method) covers the whole of the period. Obviously, we can calculate the cyclical or seasonal variations from the trend as we did in the moving average method, but equally, extrapolation from the trend which we may perform is also open to the same dangers. Neither of the two methods is 'better' than the other. The least squares method gives a straight line and shows us an overall, single figure rate of change (*b*). The longer the period of our original data, perhaps the less useful would such a single figure be, if we are dealing in terms of economic or social data which are affected by *different kinds* of influences throughout time. A straight line suggests a limited set of the same influences acting together in a single direction, which may not be the case.

More regression by computer

MINITAB, SPSS^x and SPSS/PC+ have sophisticated time series commands and modules (ARIMA) that are beyond the scope of this text. We concentrate here on how to produce simple regression on spreadsheets, SPSS/PC+ and SPSS^x. For these three packages we also develop file handling commands.

SPSS^x system files

We saw in Chapter 7 how MINITAB could save combinations of data plus data definitions and labels as a *workfile*. The same facility is available in SPSS^x as an SPSS^x *system file*. A system file stores data plus the information contained on commands like DATA LIST, MISSING VALUES, VARIABLE LABELS, VALUE LABELS. In Chapter 10 we might have wanted to save the products data plus definitions and labels in the system file PRODUCT. We could have done this by inserting one more command, SAVE, into our SPSS^x *job file*. This command should come before FINISH. In many mainframe installations, SAVE will have to be preceded by a FILE HANDLE command:

```
FILE HANDLE    SYSDATA / NAME = 'PRODUCT'
SAVE           OUTFILE = SYSDATA
FINISH
```

FILE HANDLE gives the system file PRODUCT an internal 'handle' or temporary name for the purposes of this SPSS^x job. We have chosen the internal name (or handle) 'SYSDATA'. The SAVE command is followed by the OUTFILE subcommand. The specified handle (SYSDATA) must appear on both the FILE HANDLE and SAVE commands. The above command sequence will work on a PRIME computer. Details will vary between computer install-ations.

At our next SPSS[x] session we recall the PRODUCT system file with no need to repeat DATA LIST, MISSING VALUES or the label commands. The needed commands to recall the data into SPSS[x] are now GET (with FILE subcommand) and FILE HANDLE. A complete SPSS[x] 'job' or command file (using the INDATA 'handle') might be:

```
FILE HANDLE     INDATA / NAME = 'PRODUCT'
GET             FILE = INDATA
FREQUENCIES     VARIABLES = C2 C3
FINISH
```

This job would produce frequency tables for the two variables C2 and C3. There would be no need to re-save the PRODUCT system file. See Chapter 10 for a reminder of how to run SPSS[x] jobs.

Imagine that in a previous SPSS[x] session we have used SAVE to store the withdrawals data in Table 12.7 above as the system file 'BANK'. The two variables have been saved as 'YEAR' and 'WITHDRAW'. We now wish to perform time series by simple linear regression. The command file will take this form:

SPSS[x] regression

```
FILE HANDLE     INDATA / NAME = 'BANK'
GET             FILE = INDATA
REGRESSION      VARIABLES = YEAR WITHDRAW /
                DEPENDENT = WITHDRAW /
                METHOD = ENTER
FINISH
```

The REGRESSION command here has three subcommands. The functions of VARIABLES and DEPENDENT are clear. The METHOD subcommand here tells the package to do the simplest sort of analysis (regression techniques can become very involved). The SPSS[x] output or listing file will be complex. We can disregard everything but the constant (a) and slope coefficient (b)!

SPSS/PC+ system files are saved easily with SAVE FILE without the need for the FILE HANDLE command. In an SPSS/PC+ session the BANK system file would be created and saved in this way:

SPSS/PC+

```
DATA LIST            / YEAR 1-2, WITHDRAW 3-6.
BEGIN DATA.
0725.8
0819.2

.

END DATA.
SAVE FILE 'BANK'
FINISH.
```

Do not forget the full stop at the end of each SPSS/PC+ command! At a later session we can recall BANK and perform regression:

```
GET FILE 'BANK'
REGRESSION         VARIABLES = YEAR WITHDRAW /
                   DEPENDENT = WITHDRAW /
                   METHOD = ENTER.
```

See the SPSS[x] section above for explanations. It is the results on the last screen of output that interest us. We can exit from the package with the FINISH. command.

Spreadsheet files

Packages like Lotus 1-2-3, Works and AS-EASY-AS offer file storage and retrieval from 'menu' options. Imagine that we have entered the year (variable X) and withdrawals (variable Y) data from Table 12.7 into the Lotus 1-2-3 worksheet. The worksheet, complete with labels, can easily be saved. We press ⟨/⟩(⟨**Alt**⟩ in Works). From the pop-up menu that appears, we highlight 'File' with the cursor 'arrow' key and press ⟨**RETURN**⟩. From the next pop-up menu we highlight 'Save' ('Store' in AS-EASY-AS; 'Save as' in Works) and confirm with ⟨**RETURN**⟩. We are now asked to supply a file name. We key a file name like BANK and end with ⟨**RETURN**⟩. The package is likely to add a suffix or tag (e.g. .WK1 or .WKS) to our file name.

The process of recalling a worksheet file to the worksheet is very similar. We call up the spreadsheet and summon the menu system with ⟨/⟩. We select 'File'. From the next menu we select 'Retrieve' ('Open Existing file' in Works). We then either key the name of our previously saved file, or else we highlight the name in the file menu that has appeared, and we press ⟨**RETURN**⟩. The file contents are read into the worksheet and appear on the screen.

Spreadsheet regression

The Lotus 1-2-3 and AS-EASY-AS packages also make simple linear regression easy, again from menu options. Instructions for these two packages are virtually identical. We have read the BANK data into the Lotus 1-2-3 worksheet:

	A	B	C	D	E	F	G
1	Year	Withdraw					
2							
3	7	25.8					
4	8	19.2					
5	9	26					
6	10	27.1					
7	11	22.8					
8	12	28.7					
9	13	28.5					
10	14	24.5					
11	15	29.6					
12							
13							

Fig. 12.6 Bank withdrawals data in the Lotus 1-2-3 worksheet

To access the regression routine we press ⟨/⟩. With the cursor keys we highlight 'Data' in the menu that appears, and we press ⟨**RETURN**⟩. From the next pop-up menu we highlight and select 'Regression' with ⟨**RETURN**⟩ ('Regress' in AS-EASY-AS). In the next pop-up menu we highlight 'X-Range' ('Xdata' in AS-EASY-AS) and confirm with ⟨**RETURN**⟩.

Our task is now to tell the package which data cells form the X variable for the regression analysis. We can do this with a 'point and do' method rather than by typing commands. Move the cursor to the start of the X data in column A (cell A3). When the cursor is highlighting cell A3 *do not* press ⟨**RETURN**⟩ but simply key a full stop ⟨.⟩. Then move the cursor down column A to cell A11, the last data entry for the X variable. You will see *all* the data entries in the column become highlighted. When you reach cell A11 press ⟨**RETURN**⟩. You have now set the 'range' for the X variable, year.

It will be necessary to go through a similar process to set the 'Y-Range' for the dependent or Y variable, withdrawals. The range to select here is B3..B11.

Finally, we need to use the sub-menu within regression to highlight and select 'Output-range'. In this little routine, we simply indicate to the package where we want the regression results to appear. Having selected 'Output-range' in the menu with ⟨**RETURN**⟩ we move the cursor highlight to a convenient empty cell, e.g. B13, and press ⟨**RETURN**⟩. The regression calculation will be performed automatically in AS-EASY-AS, and by selecting 'Go' with ⟨**RETURN**⟩ from the sub-menu in Lotus 1-2-3. The results for intercept and slope appear from cell B13. Output in Lotus 1-2-3 gives us various statistics. It is the 'Constant' (a) and 'X Coefficient' (b) that interest us. In AS-EASY-AS, the output is simpler—the a coefficient of the regression equation appears as 'Intercept', the b coefficient as 'Slope'.

13

Index numbers

Method

The chapter on time series dealt with the movement and changes in a single economic series through time. Many economic happenings are not the result of merely one changing series but of a collection or group of series. The change which affects the ordinary man and woman most is certainly the change in prices of goods and services which are consumed every day. The cost of day-to-day living includes a great variety of purchases at prices which change at different rates quite frequently.

Although we can directly measure the change in price of any one type of goods, it is more difficult to measure the average change in a special group of prices. It would be extremely useful to calculate some kind of average which would tell us by how much living costs had risen or fallen. This is the job which an index number tries to perform. *An index number is a measure, over time, designed to show average changes in the price, quantity or value of a group of items*.

Most of us are chiefly concerned with retail prices, but index numbers are not restricted to these. Index numbers showing economic and business trends include *quantity indexes* (e.g. volume of industrial production, volume of foreign trade) and *value indexes* (e.g. retail sales, value of exports). Nor are *price indexes* confined to retail prices. They also include stock and share prices, raw material prices, wage rates (price of labour), etc.

Construction of an index number

PRICE INDEX

The first step in construction is to ask exactly what job the index will be expected to do. Let us assume that it should show the changes in price of a group of commonly bought goods and services. Obviously, we require as a beginning:

1. A list of commonly bought goods and services. We may specify types and varieties according to how detailed the enquiry is.
2. A list of corresponding prices of these items.

We might next ask who usually buys such goods and services. To take the individual as our 'purchasing unit' would not be satisfactory. Men, women and children make very different purchases; therefore, we would require three lists, or else have to restrict our enquiry to one of these narrow groups. It would be more useful to take the family or the household as a basic purchasing unit to include all types of purchasers.

Another question which faces us immediately is what kind of household are we dealing with—rich or poor, large or small? If we wish to cover the maximum number of people in our enquiry we should once more compromise and choose the large group of lower income households. In addition, we might choose a household of average size.

Finally, there is the time element. The index number must have a starting point. This may be an actual date, e.g. the Index of Retail Prices in which 13 January 1987 was taken as equal to 100. This starting point is known as the *Base Period* or *Base Year*.

Later comparisons are usually made on a monthly basis, i.e. the average is recalculated month by month, so as to measure the change since the base period. Some index numbers are, however, calculated on a quarterly basis, and there is no reason why an individual firm should not calculate its own internal index numbers on a weekly basis if so desired.

Supposing that, after clearing up these problems, we now add up the prices of all items in our list and divide by the number of items in order to get an average. We should certainly get a false average. Several problems would appear:

1. Does the price of 5 or 10 eggs count as one item? Should we count cheese in kilograms or grams, and should we measure milk in $\frac{1}{2}$ litres or litres?
2. If salt at 2p per 500 g rises by 50 per cent to 3p per 500 g, does this count equally with a rise of 50 per cent in the price of butter (50p to 75p per kg)? Surely, the rise in the price of butter will be felt much more by the 'basic family'?

Both these problems can be solved by the simple method of weighting. In fact, most index numbers are only weighted averages, similar to those already met with in Chapter 9. If our index is truly to represent the cost to a basic family of a group of commonly bought goods and services, then the most sensible method would be to decide, say, how much per week such a family spends, and then to split this weekly budget into expenditure on the various items.

Thus, taking only three obvious items, we might get:

Table 13.1

Item	Units	Number of units bought per week	Price per unit (p)	Expenditure (price × quantity)
Milk	Litre	12	10	120
Matches	Box	6	2	12
Coal	50 kg	2	90	180
		TOTAL EXPENDITURE		312

The weight, or importance, we attach to each item in the budget will vary according to how much of the total expenditure is spent on it. The proportion of expenditure on each item is 10:1:15 (reading down the last column). These are the weights.

Price index   By the same time next year, prices will probably have changed. We will assume that they have all risen, but that the household still uses the same *quantities* of each item. Table 13.2 shows the changes, and it can be seen that the percentage rise in price has not been the same in each case:

Table 13.2

Item	Price in Year 1 (p)	Price in Year 2 (p)	Percentage change
Milk	10	12	$\frac{12}{10} \times 100 = 120$
Matches	2	$2\frac{1}{2}$	$\frac{5}{4} \times 100 = 125$
Coal	90	99	$\frac{99}{90} \times 100 = 110$

The figures in the last column are known as *price relatives* between they show, in percentage form, the new prices in Year 2 relative to the old prices of Year 1. For instance, if the price in Year 1 is assumed to be 100, then the price in Year 2 of, say, milk would be 120. We now have sufficient data to calculate the index—the percentage change in each price, and the weights of each item.

Table 13.3

Item	Price relative	Weight	Price relative × weight
Milk	120	10	1200
Matches	125	1	125
Coal	110	15	1650
	TOTALS	26	2975

It is convenient now to work in points and we can call Year 1 the base year and count it as 100. If we were to count the price relative of each item in Year 1 as 100, and then multiply it by the weights, as above, we would obviously have $26 \times 100 = 2600$. This compares with the total price relative × weight for Year 2, of 2975. Simplifying, with Year 1 = 100, then:

$$\text{Year 2} = \frac{2975 \times 100}{2600} \ 114.4.$$

We can say that the index number has increased by 14.4 points.

Note.   We cannot say 14.4 per cent, even though we calculated in the form of percentages. Suppose the index number rose to 120 in Year 3, we should say that from Year 2 to Year 3 the index number rose by 5.6 points. It would be wrong to say that it had risen by 5.6 per cent, because 5.6 per cent of the previous figure of 114.4 is $\dfrac{5.6 \times 114.4}{100} = 6.4$, and, when this is added to 114.4, we get 120.8 as the index number for Year 3, instead of the true figure of 120. To avoid confusion, therefore, it is far better to express the index numbers themselves in units of points, which have no connections with fractions or percentages.

This method of calculating the index seems reasonable enough. Yet objections may be raised. Some might say that the increase of 14.4 points gives a false impression, because freak weather produced abnormally low prices in Year 1, and abnormally high prices in Year 2. It is easy to see that if this were true, the increase in the index from Year 1 to Year 2 would be greatly magnified. It is argued that a year in which prices are about normal should be chosen as the base year. In this way, the abnormal, temporary price movements could easily be recognized for what they are. It is easy to see, also, that if the base year was a high price year, and this was followed by a year in which prices increased normally, the index would appear to have decreased! Why should we bother to object? The reason is that many wage and salary claims (among other things) are heavily dependent on the index as an argument in favour of higher rewards. If it is possible to calculate more than one figure for the index, wage and salary earners would naturally prefer the one which suits their case. As it is impossible to suit everyone, the selection of almost any year as a base will cause argument in some quarters.

1. *Purpose of the index*: Until this vital point is decided, planning is impossible. **Summary of the steps**
2. *Nature of the index*: In a retail price index, for instance, the list of items (often called a 'representative basket of goods and services') must be decided, and the corresponding prices known for each period of calculation.
3. *Type of index*: The one described above is a 'weighted arithmetic mean' of prices, using price relatives.
4. *Unit of enquiry*: In our example, this is the household in the lower income groups. Size of household, i.e. average number of persons, may be used to define the unit even further, but in practice, with a large number of households, a random sample might be taken (see Chapter 2).
5. *Base year*: This should be a normal year for the index, so far as it is possible to choose this. For example, we should not start a price index in a year of very low prices (a slump), because all future calculations will show an exaggerated rise. Also, the pattern of expenditure at such a time (the 'weights') might not be typical, because of unemployment, short-time working, etc. The index in our example is often called a 'fixed base' index, because the first year of calculation equals 100 points, and future years are calculated from this base.

The construction of a quantity index follows exactly the same method as the **Quantity** price index. Instead of using price relatives we use, of course, quantity relatives. **index** The weights used are, once again, proportional to the expenditure on each item in Year 1. The index number of quantities calculated below shows the same items as were used for the price index:

Table 13.4

Item	Quantity in Year 1	Quantity in Year 2	Quantity relatives	Weights	Quantity relative × weight
Milk	12 litres	12 litres	$\frac{12}{12} \times 100$ = 100	10	1000
Matches	6 boxes	3 boxes	$\frac{3}{6} \times 100$ = 50	1	50
Coal	2 (50 kg)	3 (50 kg)	$\frac{3}{2} \times 100$ = 150	15	2250
			TOTALS	26	3300

If Year 1 (the base year) = 100, then the index number for Year 2 will be $\frac{3300}{26}$ = 126.9 points, an increase of 26.9 points.

It may be noticed that, in this example, only the quantity of coal increased, while the quantity of milk remained the same as in Year 1, and the quantity of matches actually decreased. However, the final index number still shows an increase because the far heavier weight given to coal easily cancels out the fall in matches with plenty to spare, even though one has increased by 50 per cent and the other has decreased by 50 per cent. An *unweighted* arithmetic mean would (quite wrongly, of course) have shown no change in the index.

Types of index

The indexes dealt with above are often known as price, or quantity, relative indexes. Many other types of index have been invented and are used. As all are averages of one kind or another, they have the advantages and disadvantages common to averages. Just as there is no perfect average, so there is no ideal index. Some are more suitable than others for particular purposes.

The aggregative index

The aggregative method of calculation will give the same results as the 'relative' method, and it can be used to calculate a price or a quantity index.

An example of the price index is given in Table 13.5, using the same material as above.

If Year 1 (the base year) = 100, then the index number for Year 2 will be $\frac{357}{312} \times 100$ = 114.4 points.

By this method there is no need to convert the prices into price relatives. The formula to remember is simply:

$$\frac{\text{Total (price [Year 2]} \times \text{quantity [Year 1])}}{\text{Total (price [Year 1]} \times \text{quantity [Year 1])}} = 100 \qquad \text{(F.13.1)}$$

Table 13.5

Item	Price in Year 1 (p)	Price in Year 2 (p)	Quantity in Year 1	Price (Year 1) × quantity (Year 1)	Price (Year 2) × quantity (Year 1)
Milk	10	12	12	120	144
Matches	2	$2\frac{1}{2}$	6	12	15
Coal	90	99	2	180	198
			TOTALS	312	357

The quantity index can easily be calculated from the amended formula:

$$\frac{\text{Total (quantity [Year 2]} \times \text{price [Year 1])}}{\text{Total (quantity [Year 1]} \times \text{price [Year 1])}} = 100 \qquad \text{(F.13.2)}$$

This is a slightly different approach from the price relative method, and the latter is usually preferred in practice.

The previous examples have all used the arithmetic mean to average the items. In theory we could use any of the other averages (geometric mean, median, or mode), and the present method uses the geometric mean.

The geometric index

Weights are employed, as before, but in this case we use the *logarithms of the prices* for Years 1 and 2, instead of the prices themselves. The rest of the calculation is similar to Table 13.5, except that we must subtract the final total for Year 1 from that for Year 2, and divide the result by the total weights. The answer is in logarithms, so we find the anti-log, and multiply the result by 100 to give the index number for Year 2.

Table 13.6

Column 1	Column 2	Column 3	Column 4	Column 5	Column 6	Column 7	Column 8
Item	Price in Year 1 (p)	Price in Year 2 (p)	Quantity in Year 1 (weights)	Logs of Year 1 prices	Logs of Year 2 prices	(Year 1 logs) × weights (Col. 5) × (Col. 4)	(Year 2 logs) × weights (Col. 6) × (Col. 4)
Milk	10	12	12	1.0000	1.0792	12.0000	12.9504
Matches	2	$2\frac{1}{2}$	6	0.3010	0.3979	1.8060	2.3874
Coal	90	99	2	1.9542	1.9956	3.9084	3.9912
TOTALS			20			17.7144	19.3290

(Column 8) − (Column 7) = 19.3290 − 17.7144 = 1.6146

$$\text{and } \frac{1.6146}{20} = 0.0807$$

Anti-log 0.0807 = 1.204

With Year 1 (the base year) = 100, then the index number for Year 2 will be $1.204 \times 100 = 120.4$ points.

The chain base

All the previous examples in this chapter have been calculated on the fixed base method, that is, by selecting a base year (100), and taking the changing prices (or quantities), as a percentage of that year. Another method is the 'chain base' method by which the changing prices (or quantities) for each year (or period) are taken as a percentage of the year immediately before. This method is suitable when weights are changing rapidly because new items are being brought into the index and old items are dropping out. Thus, over a period, if changes in weights have taken place, the chain base method would, year by year, have modified itself to take account of these. In the fixed base method, such changes in weights could not have been included in the index until they amounted to such proportions that the whole index would have to be revised to prevent its going out of date. Although the chain base method might be superior for some purposes in the fairly short period, there is bound to come a time when the index number of the last year of the series bears little serious relation to the first year of the series (e.g. if weights have been changing fairly frequently). In the table below, years are shown in row 1, a chain base index (calculated as stated above) is shown in row 2. In row 3 is shown the conversion of row 2 into a fixed base index which will allow the student to compare the two indexes:

Table 13.7

Year	1	2	3	4	5
Chain base	100	103	107	110	115
Fixed base	100	100 × 103	103 × 107	110.2 × 110	121.2 × 115
		100	100	100	100
	= 100	= 103	= 110.2	= 121.2	= 139.4

The pattern of calculation is shown by the guide arrows.

The reasoning behind the above steps is as follows:

1. In Year 2, prices had risen by three per cent $(103 - 100)$ compared with Year 1. Therefore, on a fixed base, prices in Year 2 would be $\dfrac{103}{100}$ of those in Year 1 (103).

2. In Year 3, prices had risen by seven per cent, compared with Year 2. With a fixed base, they would rise by seven per cent of the previous year's figure (103). That is, the new index is $\dfrac{107}{100}$ of $103 = 110.2$ points.

3. In Year 4, prices had risen by 10 per cent, compared with Year 3. The fixed base index is thus 10 per cent more than the previous figure of 110.2. That is, $\dfrac{110}{100}$ of 110.2, and so on.

Use and interpretation

As in the case of many published figures, members of the public are apt to put far too much faith in index numbers, and to use them for arguments for which they were never intended. It is perhaps too much to ask that anyone who uses the Index of Retail Prices to support or deny a wage claim, or to criticize or praise the Government, should know the faults and limitations of this index. Most people are only too willing to cling with blind faith to an official figure, especially if it tends to support their actions.

1. Tastes and habits change in the course of time. Therefore, weightings become out of date, together with the index. For example, there is more money spent on leisure activities than there was 50 years ago. This problem is overcome to a certain extent in the latest Index of Retail Prices by the use of a chain base. **Limitations of index numbers**
2. Inventions may make certain products cheaper (this would, of course, be reflected in a price index), but they may create new basic forms of expenditure, such as television sets, or eliminate old basic forms of expenditure, e.g. the use of the telephone for communication, instead of costly travel. These also change weightings.
3. Data used in the calculation of indexes are rarely up to date, and often incomplete (see Index of Production).
4. Indexes usually cover only a part of the field of enquiry—e.g. the Index of Retail Prices only considers the prices of over 600 goods and service items, out of thousands of possible ones. These are chosen as being typical or representative items, but this is often a matter of opinion. From 1975 the weights have been chosen on the basis of a sample of households.
5. Because no particular year can strictly be called a normal year, the index contains a further element of approximation.
6. Index numbers contain an increasing degree of error, the longer they are used without revision. If they *are* revised, it is not strictly correct to compare one series with another. For example, a later index might represent a higher level of 'happiness' than an earlier one.
7. Indexes calculated for the whole country can be seriously in error when they are applied to regions within the country—e.g. the high cost of housing in London makes living costs much dearer than for those in the North, yet the Index of Retail Prices gave the weighting of $\dfrac{175}{1000}$ in 1989 for housing for the whole country.

8. Some kinds of data are not suitable for measurement, although they could be of great importance in an index. For example, income tax and mortgage payments are not included in the Index of Retail Prices.
9. An index is an average, with the advantages and disadvantages of an average; it tells us nothing, for example, about particular prices or groups of prices.
10. Data used in the compilation of indexes are often subject to error. For example, the Index of Production is based on returns by firms; the Index of Retail Prices is based on a sample of households. The volume indexes of exports and imports rely on traders' records.

There are many more limitations of index numbers, but perhaps enough has been said to warn the student not to take the figures too literally. No sensible person would deny, however, that properly based index numbers advance considerably our knowledge of changes in business and in society.

Misuse of indexes

Something has already been said about the choosing of a base year. The student will readily appreciate how a government may choose as a base year a past year in which prices were high in order to make present prices seem low by comparison. An opposition may do just the reverse in order to argue that a government has not kept its promise of maintaining stable prices. In a similar way, the rate at which prices have changed over a period may be the subject of much argument:

Table 13.8

Example:

	Year 1 (June)	Year 2 (June)	Year 3 (June)
Index	140	150	180

Possible quotations in the Press (in Year 3):

1. 'Prices have risen by 40 points since Year 1'.
2. 'Since Year 1 prices have risen by 29 per cent'.
3. 'Up to Year 3 prices had risen by 22 per cent.'
4. 'The average increase in prices since Year 1 has been 20 per cent'.
5. 'The average rate of increase in prices since Year 1 has been 13.5 per cent.'

The range of increase here (13.5 to 40) would seem large enough to suit anyone, although stranger results than these could no doubt be achieved with a little judicious juggling of base years! Let us see how they were calculated:

1. This is a simple, true statement, expressing the increase in points.
2. This calculation is obtained by expressing the difference in points between Year 3 and Year 1, i.e. 40 points, as a percentage of Year 1, thus:

$$\frac{40}{140} \times 100 = 29 \text{ per cent, which seems reasonable enough.}$$

3. This result is obtained by expressing the difference in points between Year 3 and Year 1, i.e. 40 points, as a percentage of Year 3, thus:

$$\frac{40}{180} \times 100 = 22 \text{ per cent}$$

This is inexcusable, because one would normally take the earlier year as the base quantity, unless specially asked to do otherwise.

4. This result is obtained by taking the difference between the years, 10 and 30,

adding them and dividing by two, thus $\frac{10 + 30}{2} = 20$ points (not per cent!).

5. This result is obtained by taking the differences between the years, 10 and 30, expressing them as percentages of previous years, i.e:

$$\frac{10}{140} \times 100 = 7 \text{ per cent and } \frac{30}{150} \times 100 = 20 \text{ per cent (correct so far)}$$

and simply adding the two percentage changes and dividing by two for an average, i.e:

$$\frac{7 + 20}{2} = 13.5 \text{ per cent}$$

The correct method to obtain a *rate* of increase is to take the geometric average $\sqrt{107 \times 120} = 13.3$ per cent. This means that if 140 is increased by 13.3 per cent the result will be approximately 159, and if this figure is increased again by 13.3 the result for Year 3 will be 180 (approximately, because only one decimal place of the geometric mean was calculated). Thus, only the geometric mean can be used to calculate a *rate* of increase.

Therefore, only calculations 1 and 2 are really correct. One fact is brought out in the statements—that of the need for clear thinking when one is speaking of points and percentages.

The remainder of this chapter is devoted to a brief analysis of the Index of Retail Prices, as an example of a price index on a chain base, and the Index of Production, as an example of a quantity index on a fixed base.

PURPOSE OF THE INDEX

The Index of Retail Prices

The purpose of the index is to measure the monthly degree of change in the retail prices of goods and services for the whole field over which households distribute their expenditure.

The index is the overall responsibility of the Central Statistical Office and is published in the *Monthly Digest of Statistics*, the *Employment Gazette*, the *Annual Abstract of Statistics*, and *Economic Trends*.

NATURE OF THE INDEX

The index is calculated from prices taken on a particular Tuesday each month for the groups and weightings shown in Table 13.9; comparative figures are given for 1972 weights, thus showing the difference over 17 years to 1986. A rearrangement of some categories took place from 13 January 1987 following the recommendations of the Retail Price Index Advisory Committee.

Table 13.9

	Group	Column 1 1972 weights	Column 2 1986 weights
I	Food	251	185
II	Alcoholic drink	66	82
III	Tobacco	53	40
IV	Housing	121	153
V	Fuel and light	60	62
VI	Durable household goods	58	63
VII	Clothing and footwear	89	75
VIII	Transport and vehicles	139	157
IX	Miscellaneous goods	65	81
X	Services	52	58
XI	Meals bought and consumed outside the home	46	44
	TOTALS	1000	1000

Each of the above groups contains a number of sections—e.g. 'Food' contains 32 sections. Within each section there may be various items—e.g. Drink is divided into tea, coffee, cocoa, soft drinks.

The calculation of the weights is based upon the findings of the *Family Expenditure Survey*, which operates as follows.

The Family Expenditure Survey was started in January 1957, the first report, covering the years 1957–9, being published in October 1961. It included Northern Ireland from 1958 onwards.

The survey is at present based upon a sample of 12 805 (an effective sample of about 1 in 2000) addresses each year, so chosen that every private household in the country has an equal chance of selection, and that the sample is spread evenly over the year.

The method used is based upon a three-stage sampling procedure, with stratification, and all members of a household over the age of 16 years are asked to cooperate by keeping detailed expenditure records for 14 consecutive days with supplementary information about regular, longer-term payments (e.g. rent, electricity and insurance). They are also asked about their incomes. About two-thirds of the households approached agree to cooperate.

During 1972, the weights quoted in Table 13.9 column 1 were used, and these represented the average pattern of expenditure during the three years ending June 1971, re-priced at January 1972 prices. From 1975 the weights have been based on the Family Expenditure Survey of the previous year.

Certain items of expenditure are not included in the index. They are: income tax payments, national insurance contributions, life insurance premiums and payments to pension funds, household insurance (except building insurance), trade union subscriptions, friendly societies, hospital funds, church collections, cash gifts, pools and other betting payments, mortgage payments for house purchase and for major structural alterations.

The reasons for not including these are that many of them are highly variable and often not measurable in units, and many are in the nature of investments rather than unavoidable consumer expenditure.

There are two further indices which the student should note in connection with the Index of Retail Prices:

1. *Producer Price Indices* These, several in number, were published for the first time in August 1983 and replaced the Wholesale Price Indices. They are compiled without including value added tax but include excise duties and taxes on petrol.
2. *Tax and Price Index (TPI)* This index measures the change which would be needed in gross income for a taxpayer to buy the same amount of goods and services allowing for changes in retail prices. Thus, if income tax and employees' National Insurance contributions (examples of direct taxes) were both increased, then the TPI would rise above the Index of Retail Prices because more gross income would be required to buy the same amount of goods and services.

Published indexes of prices often give separate indexes for groups within the index. It is official policy to give index numbers rounded to the first place of decimals.

A full description of the Family Expenditure Survey is provided by the technical handbook, *Family Expenditure Survey* (HMSO, 1989).

PURPOSE OF THE INDEX

The Index of Production

The purpose of the Index of Production is to measure the changes over time in the volume (or quantity) of production of a major part of the industry of the nation. The index acts as an economic indicator of industrial activity in the economic record of the country.

The index is compiled by the Central Statistical Office in collaboration with other government departments, and is published in the *Employment Gazette*, the *Monthly Digest of Statistics*, the *Annual Abstract of Statistics*, and *Economic Trends*.

NATURE OF THE INDEX

The Index of Production is a typical example of a 'base weighted' index, i.e. one which shows the relative importance of items in the base period. This is a common type, but the student should note that weights for other indexes may be calculated at some other period of time, e.g. on the current year's importance of items. Such weights have the advantage of being more up to date and they are

Table 13.10

Industry	Weights
DIVISION 1:	
Energy and water supply	
–Coal and coke	
–Extraction of mineral oil	36
and natural gas	180
–Mineral oil processing	12
–Electricity, gas, water	
supply and nuclear fuel	81 309
DIVISION 2:	
Extraction of metals and ores other than fuels, manufacture of metals, mineral products and chemicals	
–Metals	26
–Other minerals and	
mineral products	35
–Chemicals and man-made fibres	71 132
DIVISION 3:	
Metal goods, engineering and vehicle industries	
–Engineering and allied industries	295 295
DIVISION 4:	
Other manufacturing industries	
–Food, drink and tobacco	91
–Textiles, footwear,	
clothing and leather	47
–Other manufacturing	126 264
	TOTAL 1000

Note. See page 322 for a full list of 10 Standard Industrial Classification (SIC) Industrial Divisions from which this Index of Production is extracted.

called 'current weighted'. A base weighted index is said to be of the *Laspeyre* type, whilst one with current weights is called a *Paasche* type.

An example of both Laspeyre and Paasche type indexes, using the figures of Tables 13.4 and 13.5, are given below:

1. Laspeyre price index:

$$\frac{\text{Total [price (Year 2)} \times \text{Quantity (Year 1)]}}{\text{Total [price (Year 1)} \times \text{Quantity (Year 1)]}} \times 100 \quad \text{(F.13.3)}$$

$$= \frac{(12 \times 12) + (2\frac{1}{2} \times 6) + (99 \times 2)}{(10 \times 12) + (2 \times 6) + (90 \times 2)} \times 100$$

$$= \frac{357}{312} \times 100 = 114.4 \text{ points}$$

2. Paasche price index:

$$\frac{\text{Total [price (Year 2)} \times \text{Quantity (Year 2)]}}{\text{Total [price (Year 1)} \times \text{Quantity (Year 2)]}} \times 100 \quad \text{(F.13.4)}$$

$$= \frac{(12 \times 12) + (2\frac{1}{2} \times 3) + (99 \times 3)}{(10 \times 12) + (2 \times 3) + (90 \times 3)} \times 100$$

$$= \frac{448\frac{1}{2}}{396} \times 100 = 113.3 \text{ points}$$

The difference between the base weighted Laspeyre index and the current weighted Paasche index can be seen in the two formulae F.13.3 and F.13.4. In a price index of the Paasche type, the *quantities* are of the current year. In a quantity index of the Paasche type, the *prices* are of the current year.

The numerical difference between the two price indexes given above arises because the price rise (of all three commodities) is offset by the heavy consumption (in quantity) which we have used to multiply Year 1 and Year 2 prices. If price inflation exists *and* consumption (in quantity) is rising, then the Paasche index will be less than the Laspeyre index. If price inflation exists and the consumption is falling, the position will be reversed. The student may guess how the two indexes would compare if consumption remained the same in the two years!

Little precision can be attached to any index number, they are more reliable as trend indicators. The Laspeyre index is more popular than the Paasche type as the latter has to be recalculated each year and it is generally supposed that prices will alter more than quantities consumed from year to year.

Indexes and computer spreadsheets

MINITAB remembered   At this stage in the book MINITAB users may think that they have been forgotten! Not so! MINITAB users should read the next section on spreadsheet working. The Laspeyre price index is not difficult to recalculate in MINITAB with the commands learnt in Chapter 5. You will need to create new columns of data, e.g.:

$$\text{MTB} > \text{LET C5=C4}\star\text{C1}$$

You will also need to put interim and final results into three constants, e.g.:

$$\text{MTB} > \text{LET K1=SUM(C5)}$$

The section on Quicker working in spreadsheets on page 211 will not apply to MINITAB users.

A spreadsheet index   In these next two sections we use spreadsheets to recalculate the Laspeyre price index for the data in Tables 13.4 and 13.5. We also learn how to work quickly in spreadsheets by copying cells.

Figure 13.1 shows the screen display resulting from our spreadsheet entries of data, labels, formulae and functions.

	A	B	C	D	E	F	G
		Q1	P1	Q2	P2	P2 × Q1	P1 × Q1
1							
2							
3	Milk	12	10	12	12	144	120
4	Matches	6	2	3	2.5	15	12
5	Coal	2	90	3	99	198	180
6							
7	Total					357	312
8							
9	Laspeyre P.I.						114.4230

Fig. 13.1 Index spreadsheet example (screen display)

Of course, we asked the Lotus 1-2-3 spreadsheet to perform the calculations in columns F and G. What we see in Fig. 13.1 is the screen display, including results. The formulae and functions as actually entered into the worksheet columns F and G are shown below in Fig. 13.2.

		F	G
1		P2 × Q1	P1 × Q1
2			
3	Milk	+E3*B3	+C3*B3
4	Matches	+E4*B4	+C4*B4
5	Coal	+E5*B5	+C5*B5
6			
7	Total	@SUM(F3..F5)	@SUM(G3..G5)
8			
9	Laspeyre P.I.		100*F7/G7

Fig. 13.2 Index spreadsheet example (columns F and G as entered)

In constructing the formulae of columns F and G we have made use of the 'copy' facility to reduce the amount of typing needed. For example, we keyed '+E3*B3' (without quotation marks) into cell F3 and pressed ⟨**RETURN**⟩. We then copied this cell to F4. The package automatically changed 'E3' to 'E4', and 'B3' to 'B4'. The following instructions are for Lotus 1-2-3 and AS-EASY-AS. First use the cursor key to highlight in the worksheet the cell that you want to copy *from*. To evoke the copy cell procedure, press ⟨/⟩. In the menu that appears use the cursor key to highlight 'Copy'; press ⟨**RETURN**⟩ when this is highlighted. Now press ⟨**RETURN**⟩ to confirm the copy *from* cell. Next, move the cursor to highlight the cell that you want to copy *to*. Press ⟨**RETURN**⟩ to confirm that this is the destination cell. The result of the calculation, with the cell designations changed in the hidden formula from '3' to '4' will instantly appear. Repeat the procedure to copy cell F3 to F5.

Having keyed the formula into F3 we used copy twice in order to fill G4 and G5. We also copied the formula with in-built function ('@SUM(F3..F5)') from F7 to G7. The Works equivalent of F7 would be '=SUM(F3:F5)', without quotation marks. The Laspeyre price index is finally calculated in cell G9.

In larger problems you may want to copy whole ranges of cells in Lotus 1-2-3 and AS-EASY-AS. This can easily be done at the 'copy *from*' stage by highlighting a whole series of cells with the cursor arrow keys before pressing ⟨**RETURN**⟩. The spreadsheets' 'move' facility works in a similar manner to 'copy'.

The copy procedure in Works is just a little different. In Works, first highlight the cell that is the start of the range of cells that you want to copy *from*. Pressing the ⟨**F8**⟩ key then allows you, if you wish to copy more than one cell, to highlight a block or range of cells with the cursor keys. When the cell(s) that you wish to copy *from* are highlighted, press ⟨**Alt**⟩. From the menu that appears, highlight 'Edit' and select with ⟨**RETURN**⟩. In the next sub-menu, highlight 'Copy' and select with ⟨**RETURN**⟩. Finally, in the worksheet itself, move the cursor to highlight the cell that is the first in the range that you want to copy *to*. Press ⟨**RETURN**⟩ to confirm this as your choice of destination cell or range. The copy will then be made.

Repeat the above Laspeyre example for yourself, and then use a spreadsheet or MINITAB to recalculate the corresponding Paasche price index.

Quicker working in spreadsheets

14

Theory of sampling

Method

We have seen in previous chapters that we are frequently asked to make some assessment of a *group* of items, on the basis of a sample.

Types of universe

The group from which we draw our sample is called the *universe* or *population*. This is so, even if the items or members are inanimate, or purely abstract.

A *finite* universe is one with a fixed number of items, e.g. the population of a city, or customers of a firm. We may not know the exact number, at any moment, but there are obviously limits.

An *infinite* universe is one with no theoretical limit to the number of items, e.g. the stars in the sky.

Furthermore, the items may not actually exist. Such a universe is termed *hypothetical* or *theoretical*. For example, we could take a dice and throw it, observing the result. We could repeat the experiment an infinite number of times (or until the dice wore out!) and each result would be an item in our universe of throws. But we need not actually throw the dice at all; we could consider the theoretical possibilities of such an experiment. The above example has a universe which is infinite, but hypothetical.

Universes may also differ according to whether the thing being measured is *continuous* or *discrete* (see Chapter 5 on Frequency distribution). For example, the theoretical universe of throws of a dice will possess values such as a 'one', a 'two', and so on up to six (if it is the usual six-sided type of dice), but we cannot have intermediate values such as 3.14. Therefore it is *discrete*.

On the other hand, if we were measuring the heights of females in the U.K., we could have individual items which varied (at least in theory) by infinitely small amounts—i.e. the distribution would be *continuous*.

Estimation

The central problem of sampling is how to estimate some statistical measure of the universe from the corresponding statistic of the sample. For example, we could calculate the arithmetic mean of a sample and ask ourselves whether this is a good estimate of the arithmetic mean of the universe. Equally, we could try to estimate any of its other features, such as dispersion, skewness, etc.

We use the word 'estimate' because we can never be certain that our sample will be an exact miniature copy of the universe itself. We might take a coin and toss it ten times, getting three 'heads'. This represents a proportion of three out of ten, or 30 per cent heads. If we applied this result to the universe of tosses, we should obviously be in error, since the chances of a head, with a perfect coin, are one in two, or 50 per cent—i.e. half the universe would be heads, and half tails.

A secondary problem is to decide how much confidence we can place in our estimate. In general, we do this by specifying certain limits within which the true answer can lie, and, obviously, the wider the margin of error we allow ourselves, the greater the degree of confidence we can place in our estimate.

The accuracy of our estimates will depend upon:

Taking samples

1. The method of taking the sample.
2. The size of the sample.

The method usually preferred is that of random sampling (see Chapter 2), because the errors of estimation can be calculated in such cases. This is not true of the other methods mentioned.

In random sampling, each item in the universe must stand an equal chance of being selected, and the drawing of one sample or item must not affect the chances of subsequent drawings.

The first requirement can be satisfied for a finite universe in several ways.

Numbered cards or discs can represent the items, and the sample can be drawn as in a lottery. Even so, unless the items are well mixed, some of them may lie at the bottom of the pile, and have little or no chance of selection. The difficulties increase as the universe gets larger, and an alternative is to use a set of 'random numbers' (e.g. those of L. H. C. Tippett). This consists of tables of four-figure numbers (10 400 in all) whose random quality has been confirmed by numerous experiments (see section on Use and interpretation).

Another method, where the universe consists of a card index (say of a firm's customers), is to choose a number by chance (say 17) and take every 17th card in the collection until a sample of the required size is obtained. The number must be such that all the cards are covered, preferably several times, during the selection process. If there are about 20 000 cards, and a sample of 500 is needed, we would use a number in excess of 100 (say 160). Taking every 160th card would only give some 125 items when we have run through the universe, and we would carry on to the beginning again, giving other cards a chance of selection.

If the universe is infinite (throws of a dice), we simply carry out an experiment, by throwing, say, 100 times, and regard this as our sample. Because the number which turns up each time is a matter of chance, it is obvious that all the numbers have an equal possibility of selection—provided the dice are not loaded!

The importance of the second point can easily be seen, if we take a pack of cards as our universe. Suppose we are sampling for kings; there are four such cards in the pack, so the probability of drawing one is 4/52. If we are successful, there would only be three kings remaining. Our next draw has only a 3/51 chance of success, and so on. With a small universe, we can replace the item, i.e. the card before further drawings are made. With very large universes of, say, 10 000, the effect is negligible and may be ignored.

The sample should be as large as possible. We shall show, later, that the accuracy of our estimate varies inversely with the square root of the sample size, i.e. if we want to be twice as accurate (half the error) we must make the sample four times as large.

Samples of 1000 or more are certainly large enough for most purposes, and even those of 100 are often met.

It will be assumed, in what follows, that the above-mentioned conditions are satisfied.

Sampling fluctuations

Even when samples are correctly drawn, it is found that successive samples vary, or fluctuate, and that they do so according to a definite pattern. The precise form taken will depend upon the type of universe.

We saw that certain universes are discrete, as regards the quality or statistic being investigated, and we shall consider such cases first.

The thing being studied is known as an *attribute*, and every item in the universe either possesses the attribute, or does *not*. For example, in market surveys, every individual is, say, a cheese-eater, or not. We are not concerned here with how much cheese the person eats; only with the category in which to place that person. Rolling of dice or tossing of coins, are similar cases. We either get a six, or we do not; we get a head, or we do not.

Note. All possible alternative results (as with a dice) are grouped together in the *not* category.

Sampling of attributes

The appearance in our sample of the attribute we are examining is termed a *success*. Thus, if we were sampling a universe of business firms to investigate bankruptcies, then a bankruptcy would be regarded as a success! The non-appearance of the attribute is termed a *failure*.

The *probability* (P) of an event is expressed by a decimal, usually stated to three places only. Thus, the probability of getting a head when tossing a coin is 1/2, or 0.5.

If the chances of a success are 1 in 10, this is stated as 0.1. Similarly, 0.025 would represent 25 chances in 1000. The smallest possible figure is zero. This represents an absolute impossibility. Thus, the probability that the sun will rise in the west tomorrow is 0. Similarly, the largest figure is 1. This represents an absolute certainty such as that the sun will rise tomorrow.

If we show a success by the letter p and a failure by q, then it follows that $p + q = 1$, because the successes and failures taken together cover every possible eventuality. This fact is useful when we are only told *one* of the possibilities. We can find the other by subtracting from unity.

Suppose we have a universe, in which 50 per cent of the items are successes, and 50 per cent failures (say, black and white sweets). We take a large number of random samples, each of 50 sweets, and note the number of successes (say, the white ones). We should expect, in each sample, to get 25 white sweets ($\frac{1}{2}$ of 50) but, in fact, we should get a variety of results, ranging from 15 to 35 (approximately). If we noted the number of times that each result occurred, we would have a frequency distribution, and this could be plotted as a frequency polygon.

Note. The diagram would not be a smooth curve, since the distribution is discrete—going up in steps of one success.

Table 14.1 illustrates the kind of result one would expect if 1000 such samples were taken from our population of black and white sweets.

Table 14.1 Frequency distribution of successes when p $= \frac{1}{2}$, q $= \frac{1}{2}$ for 1000 samples of 50

Number of successes	Frequency
14	0.8
15	2.0
16	4.3
17	8.8
18	16.0
19	27.0
20	41.9
21	59.9
22	79.0
23	96.0
24	108.0
25	112.6
26	108.0
27	96.0
28	79.0
29	59.9
30	41.9
31	27.0
32	16.0
33	8.8
34	4.3
35	2.0
36	0.8
TOTAL	1000.0

The distribution is shown in Fig. 14.1, and it can be seen that it is symmetrical and very similar to the normal curve. The arithmetic mean (A.M.) and standard deviation (S.D.) could be calculated from the frequency distribution, using the short method of Chapters 9 and 10, but we have a much simpler method based upon the fact that such a distribution is predictable from the *binomial theorem*, a well-known mathematical device.

Binomial distribution If the sample size is n, and the probabilities of success and failure are p and q respectively, then:

$$\text{A.M.} = np \qquad \text{(F.14.1)}$$

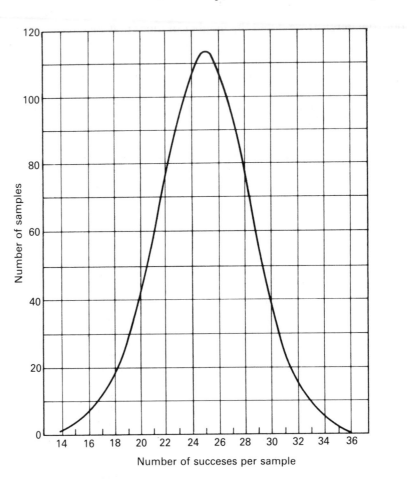

Fig. 14.1 Frequency polygon showing distribution of samples of 50 when $p = \frac{1}{2}$ and $q = \frac{1}{2}$ total samples of 1000

This is often called the 'expected' number of successes, and is the mean value of *all* samples.

For the standard deviation:

$$\text{S.D.} = \sqrt{npq} \qquad \text{(F.14.2)}$$

For example, a sample of 50 items is taken from our universe of sweets, where $p = \frac{1}{2}$ and $q = \frac{1}{2}$. Then, A.M. $= 50 \times \frac{1}{2} = 25$ white ones (or black ones, because

p and q are assumed to be equal), and S.D. $= \sqrt{50 \times \frac{1}{2} \times \frac{1}{2}} =$
(approximately).

We saw in Chapter 10 that roughly 95 per cent of the total area of th
curve is included within A.M. ± 2 S.D., and that 99 per cent is
A.M. ± 2.58 S.D. The range of A.M. ± 3 S.D. includes practically the
of the distribution (99.73 per cent).

These facts allow us to predict the likely behaviour of samples, with some
confidence.

Hence, although sample results will fluctuate, we can be 95 per cent sure that
they will lie within the range of A.M. ± 2 S.D; or, to put it another way, the
chances of a sample falling *outside* these limits are 5 in 100 (5 per cent).

This is known as the 95 per cent confidence limit, or the 5 per cent level of
significance.

Similarly, A.M. ± 2.58 S.D. is the 99 per cent limit, or the 1 per cent level of
significance.

The range of ± 3 S.D. will only be exceeded once in 500 times; it is the 0.2 per
cent level of significance.

These limits give the probabilities of exceeding a given range in either
direction (plus or minus). If we only consider one of them, then they are
halved—i.e. the chances of getting a sample with more than A.M. ± 3 S.D. (36
white sweets in our example) are 1 in 1000.

In the above example we can now say:

1. Ninety-five per cent of our samples will lie within the range
   $25 \pm (2 \times 3\frac{1}{2}) = 25 \pm 7$ white sweets.
2. The chances of getting a sample with more than 36 white ones $25 \pm (3 \times 3\frac{1}{2})$
   is only about 1 in 1000—i.e. it is so unlikely as to be rejected.

Therefore, if we do get such a sample (unless there has been some change in
the conditions of sampling, etc.), we can assume that our universe has changed,
and now contains more than 50 per cent of white sweets. This is the basic idea of
quality control (see Chapter 17).

We may also wish to find the probability of getting some specified number of
successes. This involves the use of *factorial numbers*. Factorial
$5 = 5 \times 4 \times 3 \times 2 \times 1 = 120$ and is written 5! That is, we multiply our number
by all the lesser numbers down to unity. The 1 may, in practice, be omitted,
since it does not affect the result.

Note. Factorial 0 is treated as equal to 1, although it is difficult to visualize such
an operation.

The formula we use is:

$$P_x = \frac{n!}{(X!)(n - X)!} p^X q^{n-X} \qquad \text{(F.14.3)}$$

where n = sample size, and X = our desired number of successes. For example, in samples of 10, with our black and white sweets ($p = \frac{1}{2}$, $q = \frac{1}{2}$), what is the probability of getting a sample with three white ones? Since $X = 3$, $n = 10$ and $(n - X) = 7$, we have:

$$P_3 = \frac{10!}{3!7!} \left(\tfrac{1}{2}\right)^3 \left(\tfrac{1}{2}\right)^7$$

Since p and q are equal, we can rewrite:

$$\left(\tfrac{1}{2}\right)^3 \times \left(\tfrac{1}{2}\right)^7 \text{ as } \left(\tfrac{1}{2}\right)^{10}$$

We can cancel out the whole of 7! with part of the numerator, by striking out all the numbers from 7 downwards. In fact, we need not write them in. We then cancel out the 3! into the rest of the numerator.

Note. It is usual to put dots between the numbers instead of writing out a lot of multiplication signs.

Therefore we have:

$$P_3 = \frac{10.\cancel{9}.\cancel{8}.\cancel{7.6.5.4.3.2.1}}{(\cancel{3}.\cancel{2}.1)(\cancel{7.6.5.4.3.2.1})} \left(\tfrac{1}{2}\right)^{10}$$

$$= 120\left(\tfrac{1}{2}\right)^{10} = 120/1024 = 0.1171$$

This means that the chances are 117.1 in 1000 trials that we should get exactly three white sweets, or alternatively, if we took 1024 samples, 120 of them would have three successes. The student should note that a *probability* (as above) can always be converted into a *frequency* by multiplying it by the total number of samples (total frequencies).

Note. Although the expected frequency (in samples of 10) is five white sweets, in roughly 12 per cent of the samples, we would actually get only three white ones. This illustrates the effect of sampling fluctuations. It should also be remembered that our samples here are small ones, and this makes for wider fluctuations.

Example 1:

In a random sample of 100 people, 40 per cent of them eat cheese. What proportion of the whole population are cheese-eaters?

Here we have $p = 40/100 = 0.4$; $q = 0.6$; $n = 100$. Therefore:

$$\text{A.M.} = np = 40. \quad \text{S.D.} = \sqrt{npq} = \sqrt{100 \times 0.4 \times 0.6}$$

$$= \sqrt{24} = 4.899$$

Whatever the actual proportion in the population, it should be within ±3 × 4.899 = 14.697, of the expected frequency of 40 per 100.

Therefore, the true proportion is 40 ± 14.7 per cent (approximately), i.e. between 54.7 and 25.3 per cent.

If we had chosen the 95 per cent level (2 S.D.), this would have given us a possible variation of 2 × 4.899 = 9.798 in each 100 (say, 9.8 per cent). Therefore, we could be 95 per cent sure that the true proportion was 40 ± 9.8 per cent.

The reader might object that a possible variation of nearly 15 per cent is too wide to be of any value. In this case, the remedy is to increase the sample size, say to 1600. If the same result of 40 per cent were obtained, we would now have:

$$\text{A.M.} = np = 1600 \times 0.4 = 640 \text{ cheese-eaters}$$

$$\text{S.D.} = \sqrt{npq} = \sqrt{1600 \times 0.4 \times 0.6} = \sqrt{16 \times 24} = 4\sqrt{24}$$

This is four times the original standard deviation, and equals 19.596. Taking a possible variation of 3 × S.D. gives us 58.788, but this is in relation to a value of 40 per cent of our larger sample, i.e. 640.

We now have a result of 640 ± 58.788, which in percentage form is 40 ± 3.676 per cent (58.788 out of 1600).

Note 1. This is only a quarter of the original variation (14.7 per cent), because our sample is 16 times bigger. That is, the error is inversely proportional to the square root of the sample size, and $\sqrt{16} = 4$.

Note 2. The theory assumes that p and q are the proportions in the universe itself, but if these are unknown (as in this example), no great error arises from using the sample figures, provided the sample is sufficiently large.

Note 3. When p and q are not equal, the sampling distribution becomes skewed. The greater the inequality, the greater the skewness, but increasing the sample size will offset this tendency to a great extent, and we can still use our confidence limits.

Note 4. Instead of working on the *number* of successes in our samples, we could use the *proportion*, that is, 1/nth of the number in the sample. The amended formulae now are:

$$\text{A.M.} = p \qquad\qquad \text{(F.14.4)}$$

$$\text{S.D.} = \sqrt{\frac{pq}{n}} \qquad\qquad \text{(F.14.5)}$$

Example 2:
A sample of 1000 days from the weather records of a city shows that 10 per cent of them are frosty. What is the actual proportion of frosty days?

Here $p = 1/10$, $q = 9/10$, and S.D. $= \sqrt{1/10 \times 9/10 \times 1/1000} = \sqrt{9/100\,000} = 0.0095 = 0.95$ per cent.

Taking 3 × S.D. gives us 2.85 per cent, and our estimate is 10 per cent ± 2.85 per cent.

The Poisson distribution

When p becomes very small (theoretically, say, 1 in 1000, or less) but the sample size is very large (in theory, over 1000), we get a special case, which is best described by the amended formulae:

$$\text{A.M.} = np \qquad \text{(F.14.6)}$$

$$\text{S.D.} = \sqrt{np} \qquad \text{(F.14.7)}$$

This type of distribution (known as the *Poisson*) is found where we are sampling for a comparatively rare event, e.g. the chances of having a particular type of accident or catching some rare disease (see also section on Use and interpretation).

The $3 \times$ S.D. rule may still be applied, but the other limits should be used with caution because of the possible skewness of the distribution.

Example 3:

A sample of 4000 workers in an industry shows that four of them suffer from a particular industrial disease. What is the estimated proportion in the industry as a whole?

Since $n = 4000$, and p is 4/4000, i.e. 1/1000, we have A.M. $= np = 4000 \times 1/1000 = 4$; S.D. $= \sqrt{4} = 2$; $3 \times 2 = 6$. Therefore, our estimate is 0.1 per cent $\pm$ 0.15 per cent (6 in 4000 = 0.15 per cent).

In this case, the application of the minus sign gives a theoretical limit to the left of the mean of -0.05 per cent which is, of course, impossible. This is because of the skewed nature of the distribution. The practical limits of the range are therefore 0 to 0.25 per cent.

Our previous formula for estimating the probability of getting a specified number of successes also requires modification, and now becomes:

$$P_X = \frac{(m)^x}{X!} \times \frac{1}{(2.718)^m} \qquad \text{(F.14.8)}$$

where $m = $ A.M. (i.e. np) and 2.718 is a *constant* (like π in the formula for the area of a circle).

A practical example will make clear the use of this last formula.

Example 4:

In a mass production process, where 0.1 per cent are defective, the articles are packed in boxes of 1000.

1. What proportion of the boxes would we expect to be free from defective articles?
2. What proportion would contain *two or more* defectives?

Here we have $n = 1000$ and $p = 0.001$ (or 1/1000).
Therefore $m\ (= np) = 1000 \times 1/1000 = 1$ defective.

1. Applying our formula for $X = 0$, we have:

$$P_0 = \frac{(1)^0}{0!} \times \frac{1}{(2.718)^1}$$

Since any number to the power of $0 = 1$, and remembering that $0!$ also equals 1, the first fraction reduces to unity, and can be ignored.

We therefore have $\frac{1}{2.718} = 0.3679$.

Since $P_0 = 0.3679$, we should expect no defectives 36.79 times in 100,

i.e. 37 per cent

2. When $X = 1$ we have:

$$P_1 = \frac{(1)^1}{1!} \times \frac{1}{(2.718)^1}$$

The first fraction is again unity, and the answer is again 0.3679.

Adding the two answers together gives $0.3679 + 0.3679 = 0.7358$.

This is the probability of 0 or 1 defectives. To get the probability of two or more, we subtract our previous answer from unity,

i.e. $1 - 0.7358 = 0.2642$ or 26.42 per cent

The student should be quite sure that this last step is understood, because it often arises in examinations. The explanation follows.

If we add together the successive probabilities of 0, 1, 2, 3—up to 1000 defectives per box, we have got all possibilities. This is a certainty, and in our terminology is represented by unity. If we subtract from this total, the sum of the first two probabilities (0 and 1), we are left with 2, 3, 4, . . . up to 1000, which is the same as saying 'two or more'.

So far, we have mentioned the standard deviation of the universe, and we have also talked about the standard deviation of our samples. In order to avoid confusion, it is usual to refer to the standard deviation of our samples as the *standard error*. In other words, if we consider the distribution of Fig. 14.1, which we saw was like the normal curve, then it can be described in a similar manner by mentioning two features—its average value (in this case 25) and its standard deviation. The latter measure we shall henceforth call the standard error (S.E.). **Standard error**

The probable error is the equivalent of 0.6745 S.E., and is so called because, in a symmetrical distribution, a range of A.M. $\pm$ 0.6745 $\times$ S.E. includes exactly half the items, i.e. it is as likely as not that a sample value will fall inside (or outside) this range. Hence it is, in a sense, the most probable result. **Probable error**

Apart from examination questions (where it still persists) this measure is of little importance.

Significance   It often happens that we expect a particular frequency of successes in our sample, either on the basis of our knowledge of the universe itself, or because our universe is purely theoretical and is therefore predictable on the strength of the underlying theory.

Suppose we get an actual frequency of successes which differs appreciably from expectation. If our sampling technique is above suspicion, the question arises whether the difference can be due to sampling fluctuations, in which case it might be reduced, or even eliminated, in further samples.

If the difference is too great to be explained in this way, it is said to be *significant*, i.e. it means something. What it means will depend upon circumstances. In general, it casts doubt upon our original assumptions. Either the universe has a different proportion of our attribute than we thought, or our theory is shown to be at fault. At least it directs the attention of the investigator towards the need for further enquiry and testing.

To determine what is significant, we use one or other of our confidence limits.

Example 5:

A dice is thrown 1024 times, and if a four, five, or six turns up, it is regarded as a success. The total number of successes recorded is 608. Is this result significant?

Since a dice has six sides, and three of them count as a success, the probability of a success $(p) = 3/6 = \frac{1}{2}$. Hence, $q = \frac{1}{2}$ also.

This experiment may be regarded as a sample of 1024 from an infinite universe, i.e. $n = 1024$.

Using our binomial formulae we have:

$$\text{A.M.} = np = 1024 \times \tfrac{1}{2} = 512$$

$$\text{S.D.} = \sqrt{npq} = \sqrt{1024 \times \tfrac{1}{2} \times \tfrac{1}{2}} = \sqrt{256} = 16$$

The expected frequency of successes according to our theory (the binomial theorem) is 512. Actually, we have 608, a difference of $608 - 512 = 96$. Since the standard error of our sampling distribution is 16, this represents a deviation of $96/16 = 6$ S.E. from the arithmetic mean.

We have seen that the chances of exceeding 3 S.E. are roughly 1 in 1000. In fact, a deviation of 6 S.E. would only happen 1 in 10 million times. This result is clearly significant, and suggests that our basic assumption $(p = \frac{1}{2})$ was wrong—i.e. the dice was biased, or loaded.

If our result had been 550 successes, the deviation from 512 would have been $550 - 512 = 38$. This represents just over 2 S.E. Hence, we could say the result was significant *at the five per cent level*. In other words, when judging significance we have to choose between two alternatives:

1. That a deviation is due to sampling fluctuations.
2. That it is due to something else.

If the first explanation is unlikely, we go for the second. In the latter case, if a 20 to 1 chance is regarded as unlikely (which is, to some extent, a matter of opinion) we choose the second explanation, and say that the result is significant—at that particular level.

Note. Although a significant result casts doubt upon the original assumptions, the converse does not apply. In other words, if the expected frequency is based upon some theory or hypothesis, the fact that the deviation is well within 3 S.E. does not prove that the hypothesis is correct. It only shows that there is no disagreement between the two.

Sample differences

Suppose two samples give different proportions of our attribute. Can we say that they are therefore drawn from two different universes, or might it be that the difference is due to sampling fluctuations (i.e. it is not a 'real' difference) and could disappear in the case of two other samples, taken in exactly the same circumstances? If we assume that they are both from the same universe, we need to know something about the distribution of sample differences, in such cases.

Suppose a large number of such pairs of samples were taken, and the difference between each pair was noted each time. If we take the differences in the same way throughout (e.g. sample 1 − sample 2) we should find that some differences were positive, and some negative, and we would expect the *average* difference to be zero, because we are assuming the same universe throughout, and therefore there should, in theory, be no difference.

The differences could be represented by a frequency distribution and the frequencies plotted, as in Fig. 14.1. We should get a similar type of graph, but with an arithmetic mean equal to zero, and a standard error given by the formula:

$$\text{S.E.}_{\textit{diff.}} = \sqrt{p_0 q_0 \left(\frac{1}{n_1} + \frac{1}{n_2} \right)} \qquad \text{(F.14.9)}$$

where p_0 and q_0 are the proportions given by the two samples combined, and n_1 and n_2 are the sizes of our two samples (they need not be the same).

We can now apply our levels of significance as before. If we get a difference greater than 3 S.E., this is so unlikely as to be rejected—i.e. our hypothesis that the samples were from the same universe was wrong. The difference between the two samples is a real one (significant) and cannot be explained by sampling fluctuations.

Example 6:
A sample of 400 people from Town A reveals that 40 per cent have blue eyes. A sample of 600 from Town B shows only 35 per cent with blue eyes. Is this difference significant?

We make an estimate of the proportion in the universe (assumed the same) by combining the two samples. This gives a more reliable result, because the sample size is thereby increased.

Table 14.2

	Sample size	Percentage blue-eyed	Number of blue-eyed
Town A	400	40	160
Town B	600	35	210
TOTALS	1000		370

Therefore, the percentage of blue-eyed in the two together $= 37$, i.e. $p_0 = 37$ per cent, $q_0 = 63$ per cent.

Note 1. When working in percentages $p + q = 100$ per cent (not unity).

Substituting these values in our formula, and noting that $n_1 = 400$, and $n_2 = 600$, we get:

$$S.E._{diff.} = \sqrt{37 \times 63\left(\frac{1}{400} + \frac{1}{600}\right)} = \sqrt{37 \times 63\left(\frac{3 + 2}{1200}\right)}$$

$$= \sqrt{\frac{37 \times 63 \times 5}{1200}} = 3.117 \text{ per cent}$$

Note 2. When working in percentages, the answer is a percentage. The actual difference between our two samples is 40 per cent $-$ 35 per cent $= 5$ per cent. Since 1 S.E. $= 3.117$ per cent, the difference is $5/3.117 = 1.6$ S.E. (approximately).

This is not even significant at the five per cent level, which would require a difference of 2 S.E.—i.e. the two samples could very well be drawn from the same universe.

An alternative approach is to assume that the two samples are from *different* universes, and test whether the actual difference between them is capable of disappearing in further samples. The formula for this is:

$$S.E._{diff.} = \sqrt{\frac{p_1 q_1}{n_1} + \frac{p_2 q_2}{n_2}} \qquad \text{(F.14.10)}$$

where p_1 and $p_2 =$ the proportions in each of the two samples, and n_1 and $n_2 =$ their sizes.

$$S.E._{diff.} = \sqrt{\frac{40 \times 60}{400} + \frac{35 \times 65}{600}} = \sqrt{6 + 3.79} = \sqrt{9.97}$$

$$= 3.129 \text{ per cent}$$

Again, the difference of five per cent is under 2 S.E. and is not significant. That is, the difference might not show itself in two further samples.

Use and interpretation

1. We have already seen that an accurate estimate of a universe can be made from a sample, provided it is correctly drawn. This is obviously cheaper and more time-saving than a complete investigation of every item; e.g. in social surveys and public opinion polls.
2. Frequently the information is required quickly, or at regular intervals, so that it would not be possible to make a complete enumeration each time; e.g. in dealing with index numbers, which must be produced monthly; or in quality control, where tests are frequently made hourly.
3. Sometimes, the information needed can only be obtained by destroying the item investigated, e.g. in quality control. A manufacturer of electric lamps needs to know the burning life of the products. They cannot all be burned until they fail, or there would be no products left for sale! Destruction tests are essential for vehicle tyres and similar manufactured products, and sampling is the only way.

The student should understand the important difference between sampling *error*, which is the inevitable fluctuation of samples, and *bias*, which was described in Chapter 2.

Two particular kinds of bias are likely to be the subject of examination questions.

One is known as *non-response*, and refers to the failure of certain items in the sample to respond to the enquiry. This may be deliberate, as when a postal questionnaire is not returned by the addressee, or when a person being interviewed refuses to answer. This indicates a certain attitude, and since all such people will be excluded from the final figures, the sample is no longer truly representative.

It may also be due to chance, as when an interviewer calls at an address on the sample list, and, getting no reply, calls next door instead. The original choice might be a young couple who both work, or elderly people who are deaf or infirm. The fact that the occupants are available during the day places them in a different social group from their neighbours, and the balance of the sample is therefore upset. Sometimes, bias may be deliberately introduced for sound statistical reasons.

This occurs when the frequency of occurrence of certain items in the sample should be proportional to their *size* or *importance*, rather than to their *probability of occurrence* as in ordinary random sampling.

For example, a firm supplies products to a particular industry which includes a few very large concerns and a large number of small ones. It is desired to make a sample survey of the industry, for sales purposes. If a random sample were taken, the chances of a particular firm being included in it (i.e. its probability of occurrence) might be, say, 1 in 100. If that firm, by virtue of its size, represented 20 per cent of the industry, its importance would be 1 in 5. Consequently, a stratified sample is more useful, with a system of weights, such that the chances of inclusion of the large firms are weighted, according to their relative importance. This might be done by drawing, say, 75 per cent of the sample from such firms, or by multiplying the results from each by an appropriate weight, before totalling the sample figures.

Sample size independent of universe

We have already explained that the reliability of samples increases with their size, and this is apparent from the various formulae quoted, where the size of the sample always appears under a square root sign. The student might wonder why the size of the universe does not also enter into it.

One would assume that a sample of 100 out of a village with a population of 2000 (i.e. 1 in 20 of the inhabitants) would be more reliable than one of 1000, out of a population of 200 000 (i.e. 1 in 200). Yet none of the formulae for sampling takes any note of this. The underlying assumption in sampling theory is that the universe is very large and, in general, this is true. Even if we sample from a pack of cards, the number of possible samples is infinitely large. Provided we replace each card after drawing it, we could go on indefinitely.

Random numbers

Reference has already been made to these, and several such tables are available. We now explain their use.

Suppose we want a random sample of 1000 from the inhabitants of a town. We take any of the numbers in the table, and write them down, until we have 1000 of them. We then consult the Electoral Roll, and take the people who correspond with the numbers. For instance, if our first number were 0672, we should take the 672nd name on the list, and so on. The reader might object that certain inhabitants (e.g. children) are not on the list and have, therefore, no chance of being selected. If we wish to include them, we could visit the address in question and ask all the occupants the relevant questions.

We mentioned that these numbers are limited in size (e.g. those of L. H. C. Tippett contain four-figure numbers). If the Electoral Roll contained more than 9999 names, those at the end would have no chance of selection, because there are no corresponding numbers. We can easily overcome this difficulty by running the numbers into one another, and reading them off in groups of, say, six digits, which would cover up to 1 million of a universe.

This method is commonly employed by large firms who wish to carry out a market survey.

Sample differences

These tests of significance are very useful in order to discover whether a given universe has changed between samples. This is of major importance in quality control (see Chapter 17), but the basic principle can be used in many other business situations. We consider its use in advertising.

Example 7:
A manufacturer of breakfast cereals takes a random sample of 1000 households in a certain area of the country. It reveals that 10 per cent of them use the company's product. After an intensive advertising campaign in the area, the manufacturer takes a further sample of 1000 which reveals that 15 per cent use the product. Is the difference between the two results significant?

The manufacturer hopes that the advertising has altered the character of the universe, i.e. that the universe of the second sample is a different one from that of the first—in the sense that its liking for the product has increased. On the other hand, the difference may be simply due to sampling fluctuations.

Using formula (F.14.10), which is the easier to apply, $p_1 = 10$ per cent, $q_1 = 90$ per cent, $p_2 = 15$ per cent, $q_2 = 85$ per cent; n_1 and n_2 are both 1000, since in this case the samples are the same size. We therefore have:

$$S.E._{diff.} = \sqrt{\frac{10 \times 90}{1000} + \frac{15 \times 85}{1000}} = \sqrt{0.9 + 1.275} = \sqrt{2.175}$$

$$= 1.475 \text{ per cent}$$

The actual difference is 15 per cent − 10 per cent = 5 per cent, which is more than 3 S.E. Therefore, it is most unlikely to be due to sampling fluctuations and could not disappear if further samples were taken. It is a 'real' difference, and the result is significant.

Note 1. To find the actual difference, we merely take the smaller result from the larger. In the present case, this is sample 2 − sample 1, but since the distribution is symmetrical, we are not concerned with the *sign* of the difference (plus or minus) but only with its *magnitude*.

Note 2. The two universes are almost certainly different, but as previously explained, this does not *prove* the theory that advertising is the cause. It could be some economic or social factor, but it certainly *suggests* that the advertising has paid off.

Note 3. The student should never waste time calculating the *exact* deviation in standard errors (3.389 in the above case), but need only test whether it is more than 2 S.E. (for five per cent level), or 3 S.E., according to the chosen level of significance.

Sampling and computer packages I

We shall re-examine Example 1 above using MINITAB. In a random sample of 100 people, 40 per cent eat cheese. We want to know what proportion of the whole population are cheese-eaters. We can solve this problem with the MINITAB command CDF ('cumulative distribution function') and subcommand BINOMIAL. We do not need to enter any data into the worksheet. At the MINITAB prompt we type: **MINITAB**

```
MTB > CDF;
SUBC> BINOMIAL 100, 0.4.
```

The procedure calculates the cumulative distribution function (cumulative probabilities) for the binomial where the sample size (n) equals 100, and the sample proportion (p) with the attribute of interest (cheese-eating) equals 0.4 or 40 per cent.

The package prints a table of all cumulative probabilities. We reproduce it here in a shortened form:

```
        BINOMIAL WITH N = 100 P = 0.400
              K P(X LESS OR = K)
          21                  0.0000
           .                    .
          25                  0.0012
          26                  0.0024
           .                    .
          30                  0.0248
          31                  0.0398
           .                    .
          40                  0.5433
           .                    .
          49                  0.9729
          50                  0.9832
           .                    .
          54                  0.9983
          55                  0.9991
           .                    .
          59                  1.0000
```

We know that the actual population proportion should be within 3 S.D. of the sample proportion of 40 per cent. The interval 3 S.D. each side of the sample proportion covers 99.73 of the distribution of sample results. We want to identify the proportions that correspond to ± 3 S.D. Looking at just *one* side of the distribution we therefore search for the proportion that accounts for 99.73 plus half of 0.27 (99.73 + 0.135 = 99.865) per cent of the distribution of possible sample results. We note that 0.9987 falls between the 54 per cent and 55 per cent positions, i.e. 40+15. In the lower half of the distribution, the proportion that accounts for 0.00135 of the distribution falls between 26 and 25, i.e. 40−15. We can be virtually certain that the true population proportion is 40±50 per cent. This result agrees with our original calculation.

At the 95 per cent level of confidence (±2 S.D.) we look in the upper half of the distribution for the proportion that accounts for 95 plus half of 5 (95 + 2.5 = 97.5) per cent. We see that 0.975 falls between 49 (0.9729) and 50 (0.9832), i.e. 40+10. In the lower half of the distribution, the proportion that accounts for 0.025 of the distribution falls between 31 and 30, i.e. 40−10. Again, this result repeats that of our first working.

SPSS/PC+   Using SPSS/PC+ we shall tackle Example 1 above in a slightly different way using the procedure NPAR TESTS ('non-parametric tests'). We have the data on cheese-eating for 100 respondents. Cheese-eaters are coded as '1' and non-eaters as '0', in the variable 'CHEESE'.

In this case we shall test the hypothesis that the true population proportion of cheese-eaters is 50 per cent of 0.50. How likely is it that we should get a sample proportion of 40 per cent cheese-eaters (from $n=100$) if the actual population figure is 50 per cent?

Our SPSS/PC+ run is as follows:

```
DATA LIST        / CHEESE 1.
BEGIN DATA.
1
0
.
.
.
1
0
END DATA.
NPAR TESTS       / BINOMIAL (0.5) CHEESE.
```

Note that NPAR TESTS is followed by the BINOMIAL subcommand which specifies the test proportion of 0.5 on the variable CHEESE. All commands end with a full stop.

The SPSS/PC+ output is as follows:

```
  − − − − − Binomial Test

     CHEESE

        Cases

                              Test Prop.  = .5000
             60    = 0
             40    = 1        Obs.  Prop.  = .6000

                              Z Approximation
            100    Total      2-tailed P = .0574
```

We see that about five per cent (5.74) of all sample results would estimate the cheese-eating percentage as 40 if the actual population percentage was as much as 10 points higher (50 per cent). This result is broadly in line with the two previous workings of the example where we calculated that at the 95 per cent confidence level (i.e. as estimated by all but five per cent of samples) the true population percentage was 40 ± 10 percentage points, i.e. 40 ± 2 S.D.

The SPSS[x] equivalent of the NPAR TESTS command for *Example 1* would be: **SPSS[x]**

```
NPAR TESTS       BINOMIAL (0.5) = CHEESE (0,1) /
```

Since the proportion 0.5 is the 'default' figure in SPSS[x] and SPSS/PC+ BINOMIAL tests, the '(0.5)' could be omitted. In the SPSS[x] version of the command, we specify the two values that the variable CHEESE can take (0, 1).

15

Probability

Method

The meaning of probability

This concept is basic to many business problems and plays an important role in statistics, particularly in sampling theory (see Chapter 14). Probabilities may be regarded as *relative frequencies*, i.e. the proportion of the time an event takes place.

This relative frequency, considered *in the long run*, is the probability that the particular event will happen.

When we say 'the probability that the contract will be completed on time is 0.8' we mean that on the basis of past experience, 80 per cent of all similar contracts were finished on time.

The probability of getting *heads* when we toss a perfect coin is $\frac{1}{2}$. This does *not* mean that if we toss the coin 10 times we shall necessarily get 5 heads, but if we repeat the experiment a large number of times, we would expect to approach 50 per cent heads, and the greater the number of times we tossed the coin, the closer our approximation would get.

We can see from this that probability is a substitute for certainty. If businessmen and women could always be certain that their decisions would turn out to be correct, there would be no problem. In a world full of uncertainty, and with many alternative decisions possible, they need to know the relative probabilities of success, so that they may choose that course of action which offers the best chance. This will not protect them from expensive mistakes but in the long run they will make fewer of them.

Rules of probability

MUTUALLY EXCLUSIVE EVENTS

Two or more events are said to be *mutually exclusive* if the occurrence of any one of them excludes the occurrence of *all* the others, i.e. only one can happen.

If three events, A, B, and C, are mutually exclusive, and P = probability of occurrence, then $P(A \text{ or } B \text{ or } C)$ is given by the *special rule of addition*:

$$P(A \text{ or } B \text{ or } C) = P(A) + P(B) + P(C) \tag{F.15.1}$$

i.e. the answer is the sum of the individual probabilities.

Example 1:
The probability that a firm will move its head office to Town A is 0.3, to Town B is 0.1, and to Town C is 0.4. The probability that it will move to *one or other* of them is therefore:

$$0.3 + 0.1 + 0.4 = 0.8$$

INDEPENDENT EVENTS

Two or more events are said to be *independent* if the occurrence or non-occurrence of one of them in no way affects the occurrence or non-occurrence of the others.

If A and B represent the result 'heads' in two successive tosses of a coin, then the events are independent, since the second loss cannot possibly be influenced by what happened before.

This principle is seldom grasped by gamblers, who imagine that because a coin has come up heads, say, five times in succession, it is more likely to come up tails on the sixth trial.

On the other hand, if A stands for Mr Jones being promoted, and B for his getting a car, these events are *not* necessarily independent.

When two events, A and B, are independent, the probability that they will *both* occur is given by the *special rule of multiplication*:

$$P(A \text{ and } B) = P(A) \times P(B) \qquad \text{(F.15.2)}$$

i.e. the answer is the product of the individual probabilities, no matter how many of them there may be.

Example 2:
Miss Smith is up for interview for a new job. The probability that she will get it is 0.8, and the probability that it will rain on that day is 0.3.

Since the two events are clearly independent, then the probability that they will both happen is:

$$(0.8)(0.3) = 0.24$$

CONDITIONAL EVENTS

Two or more events are said to be *conditional* when the probability that event B takes place is subject to the proviso that A has taken place, and so on. This is usually written $P_A(B)$.
Note. $P_B(A)$ would be the other way round, the assumption being that B has first taken place.

If a firm were to spend a sum of money on more efficient machinery, it is probable that profits would increase. In this case $P_A(B)$ would indicate the probability of increased profits, provided that the new machinery were purchased. Similarly, $P_B(A)$ would be the probability that new machinery would be bought, provided that profits had *already* increased. This, of course, is also a possibility.

Events such as these are clearly not independent, and the inclusion of such possibilities enables us to re-state our previous rules in more general terms.

(*a*) *General rule of addition*

$$P(A \text{ or } B) = P(A) + P(B) - P(A \text{ and } B) \qquad (F.15.3)$$

The reason for subtracting $P(A \text{ and } B)$ is because A and B are no longer necessarily mutually exclusive, and therefore *both* events might occur.

If we used the special rule of addition, we would be counting such cases *twice*, since they are A's and also B's. We must therefore deduct the proportion of such cases in arriving at our answer.

Example 3:

Statistics show that, in a certain line of business, the probability of a firm failing through shortage of capital is 0.6 and the probability of failing through shortage of orders is 0.5.

If we used the special rule F.15.1 we should get the impossible result:

$$P(A \text{ or } B) = 0.6 + 0.5 = 1.1$$

This is impossible because the largest possible answer is *unity*. As we say in Chapter 14, this represents an absolute certainty and we cannot have a stronger probability than this.

The two possibilities are not mutually exclusive; a firm could fail through *both* these causes. If the proportion of such cases were 30 per cent (0.3), then the use of F.15.3 would give:

$$P(A \text{ or } B) = 0.6 + 0.5 - 0.3 = 0.8$$

(*b*) *General rule of multiplication*

$$P(A \text{ and } B) = P(A) \times P_A(B) \qquad (F.15.4)$$

or
$$P(A \text{ and } B) = P(B) \times P_B(A)$$

depending upon the order in which the events are assumed to occur.

Example 4:

The probability that a firm will spend an extra £10 000 on advertising is 0.6. The probability that sales will increase *provided* that the extra £10 000 is spent on advertising is 0.9. The probability of *both* events happening is:

$$P(A \text{ and } B) = P(A) \times P_A(B) = (0.6)(0.9) = 0.54$$

(c) *Conditional probability under dependence*
In *Example 4*, we assumed that the conditional probability was 0.9. If we wish to calculate such probabilities, we can use the formula:

$$P_A(B) = \frac{P(AB)}{P(A)} \tag{F.15.5}$$

Example 5:
On a Board of ten directors, four are rich (over £20 000 p.a.) and six are not; three of the rich and two of the non-rich are bankers (five bankers in all).

   Since being a banker affects the probability of being rich, the two events are dependent.

   Let B = banker and A = rich, then $P(AB) = 0.3$ and $P(A) = 0.4$ (from the data), and:

$$P_A(B) = \frac{0.3}{0.4} = 0.75$$

i.e. granted that a director is rich, the chances are three out of four that he or she is a banker.

   Similarly:

$$P_B(A) = \frac{P(AB)}{P(B)} = \frac{0.3}{0.5} = 0.6$$

i.e. granted that a director is a banker, the chances are three out of five that he or she is rich.

   If we wanted the probability of getting a non-banker, granted that a director is rich, then A = rich, as before, and B = non-banker.

   $P(AB) = 0.1$ (there is only one rich non-banker) and:

$$P_A(B) = \frac{0.1}{0.4} = 0.25$$

Note 1. To get the probability of AB we *do not* use the formula $P(AB) = P(A) \times P(B)$ since this assumes statistical independence, which is not the case.

Note 2. The probabilities 0.75 and 0.25 add up to unity. This is because *all* the rich directors are either bankers or non-bankers and the two together represent a certainty, which is a probability of one. Knowing this, we could have deduced the last answer by subtracting the first one from unity, $(1 - 0.75 = 0.25)$.

Revising first estimates of probability

The ability to revise our estimates of probability as more information comes in makes our theory of great value when a businessman or woman has to make decisions under uncertainty.

Example 6:

A manufacturer has a chemical plant which mass-produces a compound. If the plant is properly adjusted, it will produce 90 per cent acceptable batches. If incorrectly set up, it will only produce 50 per cent acceptable batches. Past experience reveals that adjustments are correctly done 60 per cent of the time.

Before we inspect the first batch produced, we can only infer that the probability of a correctly adjusted plant is 0.6, based upon past experience.

Suppose we run off four batches, test them, and find them to be correct. What is our revised probability that the plant is correctly adjusted?

We first construct the following table:

Table 15.1

1	2	3	4	5
Event (adjustment)	(P) event	(P) one good batch for event	(P) four good batches for event	Joint P of event and four good batches
Correct	0.6	0.9	0.656	0.3936
Incorrect	0.4	0.5	0.0625	0.025
				0.4186

Note 1. The probability (P) of an incorrect adjustment is $1 - 0.6$, since the two probabilities must add up to unity.

Note 2. The probability of four good batches with a correct adjustment is:

$$0.9 \times 0.9 \times 0.9 \times 0.9 = 0.656$$

Note 3. The probability of four good batches with an incorrect adjustment is:

$$0.5 \times 0.5 \times 0.5 \times 0.5 = 0.0625$$

Note 4. Column 5 is the probability of the joint occurrence of columns 1 and 4. The calculation is:

Correct $0.656 \times 0.6 = 0.3936$
Incorrect $0.0625 \times 0.4 = 0.025$
(based upon F.15.4)

The revised probability is obtained from F.15.5:

$$\frac{P \text{ of correct adjustment}}{\text{(granted four good batches)}} = \frac{P(\text{correct adjustment with four good batches})}{P(\text{four good batches})}$$

$$= \frac{0.3936}{0.4186} = 0.9404$$

i.e. 94.04 per cent sure.

We have been able to improve our original estimate of a correct adjustment (0.6) to one of 0.94, on the strength of four batches produced. In other words, we are now 94 per cent sure that the plant is correctly set up.

Of course, the four batches might not all be acceptable. For example, we might find three good and one bad but, whatever the outcome, we can still calculate the probability of such a result, using the binomial formulae of Chapter 14, and applying the same technique.

The probability of getting three good batches out of four, when the plant is correctly adjusted, is in fact 0.2916. The student might care to check this, using formula (F.14.3) of Chapter 14. This principle, of continually revising our estimates as additional results come in, is a form of *sequential sampling*.

Use and interpretation

Students often have difficulty in deciding between mutually exclusive and independent events. They should remember that mutual exclusion occurs when there are several possible outcomes of a *single* action, such as the site for a new branch.

Independence is found when there are *several* actions which can occur *together*, such as engaging a sales manager and buying some new machinery. The differences between the *general* and the *special* rules are as follows:

1. *Addition*   The distinction rests upon the interpretation of 'or' in 'A or B' which may be *inclusive* and include the possibility of *both* events. In this case, we use the *general* rule:

$$P(A \text{ or } B) = P(A) + P(B) - P(A \text{ and } B)$$

   When the events are *mutually exclusive*, then, by definition, they cannot *both* occur, so that $P(A \text{ and } B) = 0$. Consequently, F.15.3 reduces to F.15.1 which is the *special* rule of addition.

2. *Multiplication*   The *general* rule is $P(A \text{ and } B) = P_A(B)$, but if the two events are *independent*, then $P(B)$ is in no way affected by $P(A)$.

   In other words, $P_A(B)$ is simply $P(B)$. Consequently, F.15.4 reduces to F.15.2, which is the *special* rule of multiplication.

Frequency distributions

The probability for any particular class is obtained by dividing the number of items in that class by the total frequencies.

Table 15.2 Sales analysis of a firm

1	2	3
Value of articles sold (£)	Number of days	Probability (P)
100 up to 120	40	0.16
120 up to 140	64	0.256
140 up to 160	80	0.32
160 up to 200	30	0.12
200 up to 300	20	0.08
300 up to 500	16	0.064
TOTALS	250	1.000

To find the probability of selling £100–£120 on any particular day, we divide the frequency of occurrence (40) by the total frequencies (250). This gives 40/250 = 0.16, and similarly with the rest.

Note 1. The total of all the probabilities is *unity*, since the table is assumed to cover all possible sales outcomes.

Note 2. A table based upon columns 1 and 3 is called a *probability distribution*, and would be used by the manager for the type of calculation we have been discussing earlier.

Note 3. The various rules of this chapter could be applied to these probabilities, e.g. the probability that sales on any one day would be £200 or more is 0.08 + 0.064 = 0.144.

We use here F.15.1 since we might sell £200–£300 *or* £300–£500, but we cannot have *both* situations on any single day. The events are mutually exclusive.

This type of calculation would be of great value to a manager in deciding the level of stock to be carried.

Probability distributions

The figures of Table 15.2 represent a continuous type of distribution, since the sales can attain any value between (say) £300 and £500. Other situations might result in a discrete type of distribution, such as Table 14.1 of Chapter 14, which gives probabilities when we divide each frequency by 1000. In this case, we might find, say, 20 white sweets in a sample, or possibly 21, but we could not have 20.3 white ones.

All the theoretical distributions discussed in the chapters on sampling are, in fact, probability distributions. The binomial and Poisson types are discrete, and the normal distribution is continuous. These theoretical distributions are used in business situations whenever we have events that occur in a random manner, i.e. they depend upon a combination of causes which are governed by chance.

A Table such as 15.2 is a sample of 250 days taken from a much larger universe of thousands of days, assuming that the firm has been in existence for some years. As such, it is subject to sampling fluctuations, and the probabilities actually calculated from it would not agree exactly with the theoretical probabilities calculated on the basis of, say, a normal distribution. The manager, faced with a probability distribution based upon historical records would first *assume* that a particular theoretical distribution applied to the business, the choice being based upon whether the information was discrete or continuous. The manager would then calculate the theoretical frequencies predicted by it and compare them with the actual results.

This can be done by multiplying each probability by the total frequencies of the distribution.

For example, in Table 15.2, the probability for the first group is 0.16. Multiplying this by the total of 250 gives 40, which is the appropriate frequency.

A comparison of these theoretical frequencies with those actually extracted from the manager's records would reveal whether the original choice of theoretical distribution was a good one, bearing in mind that, quite apart from sampling fluctuations, *no* theoretical distribution will fit exactly a particular business situation. The manager is satisfied if the 'fit' is good enough to make reasonable deductions from it.

A more exact test would be provided by χ^2 (see *Example 2* of Chapter 17). Once a particular pattern of events has been established, the manager can then use the theory to forecast the likelihood of any given result in the future.

The probability approach to decision-making under conditions of uncertainty can be used in a variety of ways. The past behaviour of sales or production, etc., is analysed, and probabilities assigned to possible outcomes. The aim is to maximize the expectation of profit or to minimize the possible loss.

Mathematical expectation

Example 7:

A chain store is opening a new branch. The company estimates that the probability of success in site A is 0.8, and that, if successful, the annual profit would be £50 000. If not successful, the annual loss is estimated at £8000. On site B, the chance of success is 0.6, with a likely profit of £60 000, or a loss of £12 000. Where should the branch be located, so as to maximize the expected profit?

If no information were available about probable success, the directors would locate the branch on site B, since the difference between expected profits and losses is £48 000, but on site A it is only £42 000.

In the light of the information on probability, a table would be set up as shown in Table 15.3.

Note 1. Since the branch is bound to make either a profit or a loss, the two probabilities add up to *unity*. The probability of a loss is $1 - 0.8 = 0.2$ for site A, and 0.4 for site B. These are the probabilities used in column 6.

The probability of a branch 'breaking even' is simply a case of zero profit or loss. It is merely a question of definition.

Note 2. We see from column 7 that the branch ought to be at site A, since the expectation of profit is maximized.

Table 15.3

1	2	3	4	5	6	7
Site	P	Possible profit	Possible loss	(2) × (3) Product	(1 − P) × (4) Product	(5) − (6) Expectation
A	0.8	£50 000	£8 000	£40 000	£1600	£38 400
B	0.6	£60 000	£12 000	£36 000	£4800	£31 200

Note 3.   If no information on probability of profit or loss were available, and the directors of the firm were *confirmed optimists*, they would choose site B, since they would look on the bright side and go for the more attractive profit.

Note 4.   If they were *confirmed pessimists*, they would seek to minimize any possible loss, and would choose site A.

This example shows the importance of statistical analysis of available figures, when making decisions. Otherwise, the businessman or woman is at the mercy of 'hunches' and mental attitudes.

Conclusion   The central problem of business nowadays is the allocation of scarce resources of capital, work-force production, etc., so as to maximize the possible outcome in terms of profit or productivity.

The manager must use the tools of forecasting and statistical inference and, by a careful analysis of existing statistical information, estimate the probabilities associated with various alternative decisions.

Only in this way can a manager choose that combination of possible courses of action which will be most likely to achieve the chosen aims.

16

Sampling of variables

Method

In Chapter 14 we were only interested in whether items in a sample did, or did not, possess a certain attribute. We now consider the *extent* to which they possess it.

Some qualities do not lend themselves to this kind of study. For example, in our discussion of a universe of black and white sweets, the question does not arise; i.e. a sweet is either black or white—the question of *degree* of whiteness does not enter into it.

If, however, we consider qualities which are capable of variation, e.g. the *weight* of a box of sweets or the *dimension* of a manufactured component, it is obvious that all the items in our universe will possess this characteristic to a greater or lesser extent. Those which do not possess it (e.g. people who do not eat cheese), may be regarded as zero cases.

In general, the values of the variable in our universe, will be spread over a range, which in theory may be unlimited. In practice, we can usually assign limits based upon experience (e.g. the heights of human beings) or upon mathematics.

Furthermore, the individual items in our universe may vary by infinitely small amounts (even though we cannot measure them) so that the distribution becomes *continuous* instead of *discrete*.

Sampling distributions

If we take a large number of samples from such a universe, and calculate in each case some statistical measure (usually called a parameter) such as the arithmetic mean, we shall get a series of different results, and for each such measure, the results may be arranged in a frequency distribution and graphed.

The resultant graph will in each case resemble that of Fig. 16.1, and if the sample size is sufficiently large, it will no longer be a frequency polygon but a smooth curve.

This is clear when we remember that instead of having only some 20-odd values to plot, as in our previous graph, we shall now have as many values as we like, because the variation is continuous. In other words the group intervals of our frequency distribution can be made smaller and smaller, until the plots on our graph are touching one another.

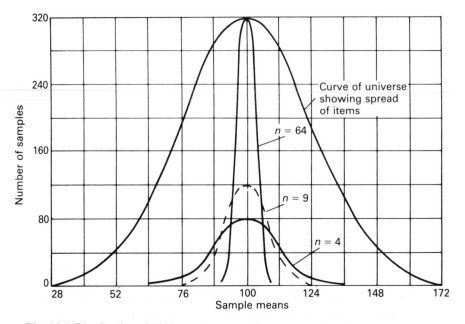

Fig. 16.1 Distribution of 1000 sample means from a normal universe (AM = 100; S.D. = 24) for various sample sizes (*n*)

To sum up, so far, we now see that if our universe is continuous, then the sampling distribution will be continuous, provided the sample size is large enough, i.e. it will be a curve.

The normal curve

We have seen (Chapter 10) that many actual universes tend to the normal form, as regards their distribution. The curves of samples from such universes will also approximate to the normal curve. In fact, even if the parent universe departs somewhat from normality, this does not seem to effect the normality of the sample curve, provided the sample size is large.

We can therefore describe these sample curves in terms of their arithmetic mean and standard error, and each statistical measure will have its own sampling curve with its own standard error. For example, if we are considering the arithmetic means of samples, then the sample curve will have a central value which corresponds with the true arithmetic mean of our universe, and the larger our sample size, the closer will be the correspondence. The standard error of our sample curve will be given by the formula:

$$S.E._{A.M.} = \frac{S.D.}{\sqrt{n}} \qquad (F.16.1)$$

where S.D. = the standard deviation of the universe from which the samples are drawn, and *n* = the sample size. If this former value is not known, we may, as before, use instead the standard deviation obtained from our sample as the best estimate available.

Note. As before, the spread of our samples is inversely proportional to the square root of the sample size (n), which appears as the denominator in our formula.

In other words, we would expect our sample averages to group themselves more closely round the real average than do the values of individual items in the universe itself.

To take an actual example, suppose we consider the heights of male adults in this country. Individual people might vary quite widely from the national average of say 1.67 m. Some might be as small as 1.30 m, while extreme cases might reach 2.15 m. But if we took samples of 100 men, it is inconceivable that any sample average would reach even 1.80 m, because some men would be below average height, while others would be above it. It is equally unlikely that any sample would have an average as low as 1.50 m. The bigger our samples, the more likely our heights will average out. Figure 16.1 illustrates the type of situation we would get.

Note. The spread of our sample means is very much less when the sample size is increased. When $n = 64$ we get a very fair approximation to the arithmetic mean of the universe, and the majority of our samples lie with ±9 units of the true arithmetic mean of the universe.

Example 1:
A random sample of 100 people in a certain city shows an average weekly consumption of meat of 700 g, with a standard deviation of 300 g. What is our estimate of the average consumption in the city as a whole?

Using formula F.16.1, we have S.D. = 300 g, n = 100.

$$\text{Hence, S.E.}_{\text{A.M.}} = \frac{300 \text{ g}}{\sqrt{100}} = 30 \text{ g.}$$

We could be 95 per cent sure that the true arithmetic mean lies within 700 g ± 2 × 30 g = 700 ± 60 g.

We could be 99.7 per cent sure that the answer lies within 700 g ± × 3 × 30 = 700 ± 90 g.

Were we to increase the sample size to 1600 then the S.E.$_{\text{A.M.}}$ would be $\frac{300 \text{ g}}{\sqrt{1\,600}} = \frac{300 \text{ g}}{40} = 7\frac{1}{2}$ g, and we could now be more precise about our estimate and say that at the 95 per cent level of confidence the true answer is 700 ± 15 g.

The normal curve is actually the graph of a mathematical equation. Suppose we had an equation of two variables, such as $Y = X^2 + 7$. We could substitute a series of values for X and the equation would tell us the corresponding value of Y in each case. We could then plot each X against the corresponding Y as a point

Use of tables of normal curve

on a graph, and if we had sufficient points we could join them up in the form of a curve.

The equation for the normal curve is much more complex than the example quoted, but the result would equally be a curve, although of a different shape. It is, of course, the familiar bell-shaped curve already referred to.

Students need not concern themselves with the equation itself, because tables are available based upon it, but they should understand the meaning and use of the tables.

In Chapter 7 we explained that in the case of a histogram the areas of the bars are proportional to the frequencies. This is equally true of our curve, and if we know what proportion of the total area of the curve is enclosed between any two perpendicular lines, we know that the same proportion of the total frequencies will be found between those two limits.

In order to standardize our table of areas, we proceed as follows:

1. We assume that the total area enclosed by the curve is unity. This will enable us to specify the *probability* of getting some particular result in our sampling, since the area of the curve covers all possible sample results, and this, according to our previous notation of probabilities, is represented by unity (a certainty).
2. We adopt as one of our perpendicular lines, the one which divides the curve into two equal parts. It will be remembered that the normal curve is unimodal and symmetrical, so this line will represent the average (arithmetic mean = median = mode) of our distribution.
3. We measure our horizontal distances from this centre line, not in any specified units but in fractions of the standard deviation of the curve. This is because the shape and area relationships of the normal curve are governed entirely by its standard deviation. It also has the merit of eliminating the units in any particular problem, so that our horizontal scale is not in kilograms or metres, but in standard deviations.
4. We measure plus deviations to the right, and minus ones to the left, of the centre line.

The standard tables tell us the probability that an item (or sample) will lie within some specified portion of the total area, but the information may be presented in a variety of ways. The commonest ones are as follows (see Figs. 16.2(a)–16.2(c):

1. *Type A*   The area enclosed between two perpendiculars at equal distances measured from the centre line; i.e. arithmetic mean ± a specified number of standard deviations. *Note.* Sometimes only the positive half is given.
2. *Type B*   The area *beyond* some specified perpendicular; i.e. the 'tail' of the curve, measured in a *positive direction* only. This tells us the probability of getting a sample in excess of some particular value, assuming it to be due to sampling fluctuations. If we wish to include negative fluctuations as well (i.e. a very small result), we simply include the other tail in addition.

3. *Type C* The whole area *to the left* of some specified vertical. This is simply the result of the unshaded area subtracted from unity.

The student should read carefully any instructions regarding the interpretation of a given table. In particular, if the largest number in it is 0.5, then it only covers *half* of the curve (see section on Use and interpretation).

The probability given in the table is always a decimal and, when multiplied by the total frequencies, gives the theoretical number of cases within the area specified. Alternatively, we may multiply our decimal by 100 and get the *percentage* of cases.

So far we have only tried to estimate the arithmetic mean of our universe from a sample. We could equally estimate any other statistic, and formulae are given for some of the more important ones in the following sections.

Estimates of universe from a sample

ESTIMATING STANDARD DEVIATION OF UNIVERSE

The formula for estimating the standard deviation of a universe from a sample is

$$\text{S.E.}_{\text{S.D.}} = \frac{S}{\sqrt{2n}} \qquad \text{(F.16.2)}$$

where S = the standard deviation of our sample, and n = the sample size, as before.

Example 2:
A manufacturer of shoelaces wishes to know the standard deviation of the company's output. A random sample of 200 has a standard deviation of 10 cm. What estimate does this provide of the true standard deviation?

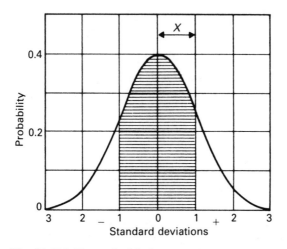

Fig. 16.2(a) Form of table for normal curve type A

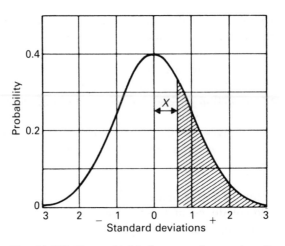

Fig. 16.2(b) Form of table for normal curve type B

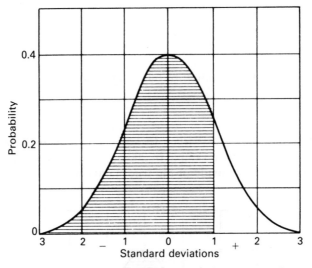

Fig. 16.2(c) Form of table for normal curve type C

Using formula F.16.2, we have:

$$\text{S.E.} = \frac{10 \text{ cm}}{\sqrt{400}} = \frac{10 \text{ cm}}{20} = 0.5 \text{ cm}$$

Hence, the true standard deviation of the company's output should lie within:

$$10 \text{ cm} \pm 3 \times 0.5 \text{ cm} = 10 \text{ cm} \pm 1.5 \text{ cm}$$

A similar problem to that of Chapter 14 now arises. How can we test whether **Sample** the difference between two sample results is significant? **differences**
Here again, we can approach the problem in two ways:

1. Assume that the samples are from the same universe, and test whether the difference between the two results could be due to sampling fluctuations.
2. Assume they are from different universes, and test whether the difference could be eliminated in further pairs of samples.

DIFFERENCES BETWEEN MEANS

The student is advised to use method 2 above, which is simpler to work out. The formula is:

$$S.E._{diff} = \sqrt{\frac{S.D._1^2}{n_1} + \frac{S.D._2^2}{n_2}} \qquad (F.16.3)$$

where $S.D._1$ and $S.D._2$ = the standard deviations of the two samples, and n_1 and n_2 = their sizes.

Example 3:
Two samples of electric lamps are taken from two different factories of a large organization. Sample A, of 100 lamps, has an average burning life of 1100 hours, with a standard deviation of 240 hours. Sample B, of 200 lamps, has an average burning life of 900 hours, with a standard deviation of 220 hours. Is the difference between the lives of the lamps significant?
Using formula F.16.3 we have $n_1 = 100$, $S.D._1 = 240$ hours, $n_2 = 200$, and $S.D._2 = 220$ hours. Hence:

$$S.E._{diff.} = \sqrt{\frac{240^2}{100} + \frac{220^2}{200}} = \sqrt{576 + 242} = \sqrt{818} = 28.6 \text{ hours}$$

The actual difference between the two means is $1100 - 900 = 200$ hours. Hence, it is well over 3 S.E. (in fact it is about 7 S.E.), so that it could not possibly be eliminated by further pairs of samples. The output from the two factories is indeed different as far as average burning life is concerned.
Note. We need not waste valuable time calculating the *exact* number of standard errors in the difference. We are concerned only with whether it exceeds our yardstick of 3 standard errors, and so on, according to our chosen level of significance.

DIFFERENCES BETWEEN SAMPLE STANDARD DEVIATIONS

A further application of the same idea is to test whether two samples, with different standard deviations, could be from the same universe; i.e. is the difference between the two results significant?

In this case, we are only concerned with the standard deviation of our samples, as related to that of the universe, and not with any other feature.

Using method 2 above, the formula is:

$$\text{S.E.}_{\text{diff.}} = \sqrt{\frac{\text{S.D.}_1^2}{2n_1} + \frac{\text{S.D.}_2^2}{2n_2}} \qquad \text{(F.16.4)}$$

Example 4:

In the previous example, we proved that there was a significant difference in the arithmetic mean life of the lamps. We now ask whether there is a significant difference between the two standard deviations. These were 240 hours for the sample of 100, and 220 hours for the sample of 200.

We now have, using formula F.16.4:

$$\text{S.E.}_{\text{diff.}} = \sqrt{\frac{240 \times 240}{2 \times 100} + \frac{220 \times 220}{2 \times 200}}$$

$$= \sqrt{288 + 121} = \sqrt{409} = 20.22 \text{ hours}$$

The actual difference between the samples is $240 - 220 = 20$ hours. This is less than one S.E. and is clearly not significant.

As a result of these two tests, we can say that the outputs of the two factories differ as regards their average burning life, but not in the variability or spread of the product.

Use and interpretation

One of the main uses of sampling techniques is in the field of quality control in industry, and we shall deal with this in detail in the next chapter.

The technique is also widely used for market research. This is a statistical investigation carried out by a manufacturer or distributor, to find out one or more of the following:

1. *Who uses the product?* It may appeal to certain income groups, or people of a particular age or sex. The occupation or interests of customers may be important, and so on.
2. *Why they use it (or not).* This may give a line on competitors, or point to certain features as important. Sometimes the answer is surprising. Reasons given in actual surveys have included the size of the lid on the container and even the dimensions of the container itself.
3. *How much people would expect to pay.* The public frequently has a built-in resistance to products outside a particular price range.
4. *Buying habits.* When and where they get it; how frequently and so on. This governs:
5. *Packaging.* The size and quantity packed. Style and design of package that appeals, etc.

6. *Opinions about the product.* What people expect or would like; criticisms, etc.

7. *Geographical differences of taste.* In the case of a firm supplying a specialized product, or appealing to a limited market, it may be possible to survey the whole of the field, but in the majority of cases a sample must be chosen. (See Chapter 2 for discussion.)

Public opinion polls are well-known applications of sampling techniques, and the student should now appreciate why different results at different times or places may have little statistical significance. This type of sample is usually a *stratified* one, so as to include a cross-section of the public.

With regard to official samples, we have already mentioned sampling in official statistics as a method of saving time and money; e.g. the 10 per cent sample in the Population Census. The Censuses of Production and Distribution also use this device, as an alternative to a full census.

The Social Survey is produced by a government department which undertakes surveys on behalf of other departments. These investigations are usually based on the method of random sampling.

The use and interpretation of the normal tables of probabilities often seem difficult to students, and we therefore give a series of examples covering all likely applications.

Example 5:

A sample of 1000 articles has an arithmetic mean of 300 g, and a standard deviation of 60 g.

1. What number of articles will lie between 270 and 330 g?

   Here we are concerned with the spread on *either side* of the arithmetic mean,

$$\text{i.e. A.M.} \pm 30 \text{ g } (300 \pm 30 \text{ g})$$

   We first reduce this to units measured in standard deviations,

$$\text{i.e. } \frac{30 \text{ g}}{60 \text{ g}} = 0.5 \text{ S.D.}$$

   We now require the shaded area of Fig. 16.2(a) where X (the distance on the horizontal scale) $= \frac{1}{2}$.

   Using the table in Appendix 1 (which is of type A but gives only *half* the area) we see that, for a distance of 0.5, the probability figure is 0.1915. We therefore *double* this, and get 0.3830. This is the *proportion* of the total area.

   Multiplying this by the total frequencies (1000) gives us 383 articles.

2. What number of articles will *exceed* 390 g?

   Here, the value of X (horizontal distance) is 390 g − 300 g = 90g (ignore the sign).

Reducing this to standard deviations, we get $\dfrac{90 \text{ g}}{60 \text{ g}} = 1.5$ S.D.

We now require the shaded area of the curve in Fig. 16.2(b).

Our table for 1.5 S.D. gives a proportion of 0.4332, but this is not the portion we want. Our table value is the area between the centre line (300 g in this example) and a vertical line at 390 g.

Remembering that the total area of our standard curve is unity, and that the right-hand portion is therefore 0.5, we subtract our tabular answer from 0.5. We now have $0.5 - 0.4332 = 0.0668$.

Multiplying this by 1000, as before, gives us our answer, namely 66.8 (67 articles).

3. What proportion of the articles will weigh *less than* 198 g?

This is $300 - 198 = 102$ g, i.e. 1.7 S.D.

This time, the distance is measured to the *left* of our centre line. The situation is similar to the shaded area of curve (Fig. 16.2(b)), except that we require the area of the *other* tail. Since the curve is symmetrical, the two tails are identical, so we can use our table directly.

For a distance of 1.7 S.D. the proportion is 0.4554. Taking this from 0.5 as before, to get the tail, we have $0.5 - 0.4554 = 0.0446$.

As a *proportion* is called for, we can let this answer stand, or multiply it by 100 = 4.46 per cent.

Example 6:

Assume a universe with a mean of 100 and a standard deviation of 20.

1. If the probability that a certain value of the variable will be exceeded is 10 per cent, what is the value?

We are told that the area of the tail in Fig. 16.2(b) is 0.1. Taking this from 0.5, to get the remainder of the right-hand portion, we have $0.5 - 0.1 = 0.4$.

We now find this value, as closely as possible, in our Table 2 (Appendix 1). The nearest value is 1.3 standard deviation (for area 0.4032).

Note. Much more detailed tables of the normal curve are available (see references).

Since one S.D. = 20, this gives us $1.3 \times 20 = 26$ (measured from the centre line = A.M. of our curve).

Hence, the value required = 100 + 26 = 126.

2. The probability that a certain *deviation from the mean* will be exceeded is 10 per cent. What is the deviation?

This type of question may be interpreted in two ways, depending upon what we mean by 'exceeded'.

(a) In the ordinary sense of the word, we mean bigger than, i.e. we need only consider the right-hand portion of the curve, and the answer is 26, as in case 1.

(b) If we mean a deviation in *either direction* ($\pm$), then we are considering the unshaded area of Fig. 16.2(a). If this is 10 per cent (0.1), then each tail is 0.05, and taking this from half the curve (0.5) gives us: $0.5 - 0.05 = 0.45$.

We cannot find this exact value in our table, but it is about half way between 1.6 S.D. and 1.7 S.D., i.e. 1.65 S.D.

Therefore the required deviation is $\pm (1.65 \times 20) = \pm 33$.

Example 7:

In a normal distribution, 30 per cent of the items are under 50, and 10 per cent are over 86. Find the arithmetic mean and standard deviation of the distribution.

1. Since 50 per cent of the curve is to the left of the arithmetic mean, we know that the vertical at 50 must also be to the left (since only 30 per cent of the curve is to the left of this line). That is, the 30 per cent is the tail of the curve.
2. The left half of the curve has an area of 0.5. The tail is 0.3 (30 per cent). Hence the area between the arithmetic mean and $50 = (0.5 - 0.3) = 0.2$. From our table, 0.2 represents a distance (X) of 0.5 S.D. Hence,

$$\text{A.M.} - 50 = 0.5 \text{ S.D.} \qquad \text{Eq. (16.1)}$$

3. By similar reasoning, the 10 per cent represents the right-hand tail of the curve ($= 0.1$).
4. Hence, the area between 86 and the arithmetic mean is $0.5 - 0.1 = 0.4$. From our table, 0.4 represents a distance of 1.3 S.D., and so:

$$86 - \text{A.M.} = 1.3 \text{ S.D.} \qquad \text{Eq. (16.2)}$$

5. We now have two equations (Eqs 16.1 and 16.2):

$$\text{A.M.} - 50 = 0.5 \text{ S.D.}$$
$$86 - \text{A.M.} = 1.3 \text{ S.D.}$$

Adding them:

$$86 - 50 = 0.5 \text{ S.D.} + 1.3 \text{ S.D.} \quad 36 = 1.8 \text{ S.D.} \quad \frac{36}{1.8} = 1 \text{ S.D., i.e. S.D.} = 20$$

6. Substitute this value in Eq. 16.1:

$$\text{A.M.} - 50 = 0.5 \times 20, \quad \text{A.M.} - 50 = 10, \text{ i.e. A.M.} = 60$$

The distribution can now be visualized as shown in Fig. 16.3.

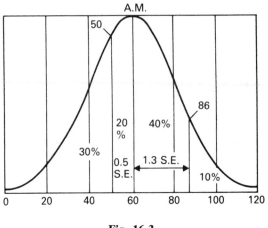

Fig. 16.3

**Determin-
ation of
sample size**

Example 8:

The standard deviation of a universe is 60 g. What size of sample must we take to be 95 per cent sure that the sample arithmetic mean will not differ from the arithmetic mean of the universe by more than 10 g?

1. We use formula F. 16.1 for the standard error of the arithmetic mean:

$$S.E._{A.M.} = \frac{S.D.}{\sqrt{n}} = \frac{60 \text{ g}}{\sqrt{n}}$$

2. We know that the 95 per cent confidence limits are:

$$A.M. \pm 1.96 \text{ S.E.}$$

This is one of the limits which the student *must* remember.
Hence:

$$1.96 \times \frac{60 \text{ g}}{\sqrt{n}} = 10 \text{ g, i.e.} \frac{117.6 \text{ g}}{\sqrt{n}} = 10 \text{ g}$$

$$10\sqrt{n} = 117.6, \ \sqrt{n} = 11.76, \ n = (11.76)^2 = 138 \text{ or } 139$$

Note. Instead of 1.96 S.E., a round figure of 2 S.E. may be used.

Sampling and computer packages II

MINITAB offers some easy-to-use commands for statistical significance tests on sample means. Our first illustration repeats *Example 1* (page 241) via the MINITAB package. We have taken a random sample of 100 people in a city. The mean weekly meat consumption is 700 g, with a standard deviation of 300 g. We wish to estimate the average consumption in the city as a whole.

The first thing to note is that the problem concerns one sample taken from a larger population. Also, as in most practical problems, we have knowledge of the *sample* standard deviation rather than the *population* standard deviation. In these circumstances, (single sample), population standard deviation unknown) the most appropriate commands within MINITAB are TINTERVAL and TTEST. The command TINTERVAL calculates a confidence interval for the true mean, while TTEST performs the related task of testing, from the sample data, a *hypothesis* about the true mean (e.g. that the true mean = 750 g).

In the original working of *Example 1*, we calculated a confidence interval. We shall do the same again with MINITAB. We therefore use TINTERVAL. The 100 observations on meat consumption have been entered as column C1 in the MINITAB worksheet. To perform the test we key (after the MTB > prompt):

```
MTB > TINTERVAL 95 C1
```

This command produces the 95 per cent confidence interval for the mean from the sample data in C1. The output that follows is:

```
        N     MEAN     STDEV     SE MEAN     95.0 PERCENT C.I.

C1     100    700.0    300.0      30.0        (640.5, 759.6)
```

Compare the results with those of the original working. Allowing for very small rounding errors they are the same—we can be 95 per cent sure that the true population mean lies between 640 g and 760 g. Try reworking this example with 95 and then with different confidence percentages—90, 99, 99.7. To rework the example you will obviously first need to input 100 sample observations into C1. We can do this with a convenient form of the SET command:

```
MTB  > SET C1
DATA > 46(400)
DATA > 4(420)
DATA > 3(980)
DATA > 985
DATA > 46(1000)
DATA > END
```

This use of SET puts into C1 46 occurrences of 400, 4 of 420, etc. The resulting fictitious dataset is highly unlikely, but is quickly input, and it gives a sample with a mean of 700 g and a standard deviation of 300 g.

The single sample command TTEST is similar to TINTERVAL but calculates a *t-test* for the variable of interest. In this example we might use the command:

MTB > TTEST 750 C1

This example calculates the probability of getting a sample mean of 700 g with a standard deviation of 300 g if the mean of the population from which the sample comes is really 750 g. Note that on the TTEST command we specify the assumed mean of the population, not the confidence percentage.

MINITAB —two samples

In *Example 3* on page 245 we have a statistical significance problem involving two sample means. The electric lamps of sample A ($n_1 = 100$) have an average life of 1100 hours and a standard deviation of 240 hours. The lamps of sample B ($n_2 = 200$) have an average life of 900 hours with a standard deviation of 220 hours. Is the difference between the lives of the lamps significant?

The MINITAB command TWOSAMPLE gives *both* a significance t-test *and* a confidence interval for the difference between two sample means. TWO-SAMPLE does for two samples the combined work of TINTERVAL and TTEST on single samples. In the illustration below, we input some highly fictitious data and use TWOSAMPLE:

```
MTB   > SET C1
DATA > 46(860), 4(880), 1320, 1325, 48(1340)
DATA > END
MTB   > SET C2
DATA > 97(680), 3(700), 3(1100), 97(1120)
DATA > END
MTB   > TWOSAMPLE 95 C1 C2
```

The output is as follows:

TWOSAMPLE T FOR C1 VS C2

	N	MEAN	STDEV	SE MEAN
C1	100	1100	240	24
C2	200	900	220	16

95 PCT CI FOR MU C1 − MU C2: (144, 257)

TTEST MU C1 = MU C2 (VS NE): T = 7.01   P = 0.0000   DF = 183

The 95 per cent confidence interval for the difference between the two sample means is 144 to 257. The interval would have to contain 0 (i.e. zero difference) for us to conclude that the two samples came from the same population. The final line of output repeats this finding, but in a different way. We test the *hypothesis* that the population mean of C1 (indicated by the Greek letter 'μ' or *Mu*) is the same as the population mean from which the sample in C2 comes. The *alternative hypothesis* is that the two population means are 'not equal' ('NE'), i.e. that the two samples come from different populations. The calculated *t-statistic* is 7.01, i.e. the difference between the two means is about 7 S.E. (compare with the original working above). It would be extraordinarily unlikely to obtain two sample means so different if the samples really came from the same underlying population.

MINITAB offers a second version of the TWOSAMPLE procedure. The command TWOT exactly reproduces the TWOSAMPLE statistics, but for data organized differently in the worksheet. We might have all 300 observations (i.e. both samples combined) in column C3 of the worksheet. Information on which sample the observation came from would have to be stored in a separate column, e.g. C4. Entries in C4 are either '1' for a sample A observation, or '2' for sample B. In the TWOT command we specify the confidence percentage (again 95 per cent), then the column of actual observations (C3), and finally the column containing the group identifier (here C4):

$$\text{MTB} > \text{TWOT } 95 \text{ C3 C4}$$

The output is the same as for TWOSAMPLE.

We return to single sample problems like *Example 1* above (meat consumption). We can use the simple and familiar CONDESCRIPTIVE command in SPSS^x (see Chapter 10). If we follow CONDESCRIPTIVE with the appropriate STATISTICS command we obtain the mean, standard deviation and standard error of the mean. We can then easily multiply the standard error by two or three to obtain the boundaries of the 95 and 99.7 per cent confidence levels (see original working of *Example 1*). Imagine that our consumption data are stored as the variable 'MEAT', and has been called into SPSS^x from a system file (see Chapter 12). The rest of our command or job file is:

SPSS^x— single samples

```
CONDESCRIPTIVE    MEAT
STATISTICS        ALL
FINISH
```

SPSS^x has the T-TEST procedure which is like TWOT in MINITAB. To solve the *Example 3* problem above (electric lamps), we use T-TEST plus two subcommands. In our SPSS^x replication we have called a system file in which the variable 'SAMP' indicates which sample the observation comes from ('1' for sample A, '2' for B). The lamp life is recorded in variable 'LIFE'. The rest of our command file is:

SPSS^x— two samples

```
T-TEST       GROUPS = SAMP / VARIABLES = LIFE
FINISH
```

253

The GROUPS subcommand indicates where the package will find the sample identifiers. The VARIABLES subcommand indicates which variable(s) the t-test is to be performed on. See the final SPSS/PC+ section below for typical output.

SPSS/ PC+ – single samples

For *Example 1* above (meat consumption) we can use the SPSS/PC+ DESCRIPTIVES command (equal to CONDESCRIPTIVE in SPSS[x]) on the variable MEAT (consumption). DESCRIPTIVES was introduced in Chapter 11. With the STATISTICS subcommand we obtain the mean, standard deviation and standard error of the mean. We can then easily multiply the standard error by two or three to obtain the boundaries of the 95 and 99.7 per cent confidence levels (see original working of *Example 1*):

DESCRIPTIVES / VARIABLES MEAT / STATISTICS ALL.

SPSS/ PC+ – two samples

SPSS/PC+ has its version of T-TEST, like SPSS[x]. See the SPSS[x] section above for an explanation of the GROUPS subcommand. In this example, the variable LIFE contains the data on lamp life, and the variable SAMP indicates whether the observation is from sample A ('1') or B ('2'). The SPSS/PC+ version of the command is:

T-TEST / GROUPS SAMP / VARIABLES LIFE.

Remember that SPSS/PC+ commands end with a full stop.

The output from the SPSS/PC+ and SPSS[x] T-TEST will be as shown in Fig. 16.4.

Independent samples of SAMP

Group 1: SAMP EQ 1   Group 2: SAMP EQ 2

t-test for: LIFE

	Number of Cases	Mean	Standard Deviation	Standard Error
Group 1	100	1100.45	240.097	24.010
Group 2	200	900.00	219.977	15.555

F Value	2-Tail Prob.	Pooled Variance Estimate			Separate Variance Estimate		
		t Value	Degrees Freedom	2-Tail Prob.	t Value	Degrees Freedom	2-Tail Prob.
1.19	0.301	7.21	298	0.000	7.01	183.46	0.000

Fig. 16.4

Ignore the 'Pooled Variance Estimate' section, plus 'Degrees of Freedom' and the 'F Value'. Our original calculations involved the safe procedure of 'Separate Variance Estimates' (i.e. *we avoided* the assumption that the two samples come from the same population). The '2-Tailed Probability' statistic gives the probability (maximum possible value of one) of discovering this much difference between the two sample means (1100–900 = 200) if there is no real difference between the means of the populations from which they are drawn. Here, the probability of discovering this much sample difference when no population mean difference exists is effectively zero (0.000)—i.e. we reject the *null hypothesis* of a single population mean and accept the *alternative hypothesis* that the samples are drawn from distinct populations.

17

Further applications of sampling

Method

The correlation coefficient (r)

In previous chapters we discussed sampling theory in general terms. We now consider a few specialized applications of these principles.

For samples from a normal universe, the standard error of the Pearsonian coefficient of correlation is given by:

$$S.E._r = \frac{1 - r^2}{\sqrt{n}} \qquad \text{(F. 17.1)}$$

where n = the sample size, as before.

For large samples (say 100 or more pairs of observations) the distribution of r approximates to the normal curve, and we can use the levels of significance previously described. Thus, a sample value which exceeds 3 S.E. is most unlikely to be due to sampling fluctuations, i.e. it is significantly different from zero, or it is 'real'.

This measure is of limited use, as will be explained in the section on 'Use and interpretation'.

Example 1:
A sample of 400 pairs of items gives a value of r of 0.3. Is there evidence of correlation?

Applying formula F. 17.1 we have:

$$S.E._r = \frac{1 - 0.3^2}{\sqrt{400}} = \frac{1 - 0.09}{20} = \frac{0.91}{20} = 0.046$$

We now apply what is sometimes termed the *null hypothesis*, i.e. we assume that there is zero correlation in the universe from which the sample is drawn, and that the value of 0.3 obtained is, therefore, due to sampling fluctuations.

If this is so, then the deviation from zero is

$$\frac{0.3}{0.046} = \text{about } 6\frac{1}{2} \text{ S.E.}$$

We have already said that a deviation of more than 3 S.E. is about the highest sampling fluctuation possible. Hence, this result is too high to be explained in this way, and our null hypothesis must be rejected. In other words, the correlation is real.

Note. If the same result had been obtained from a sample of 25, however, the calculation would be:

$$\text{S.E.}_r = \frac{0.91}{\sqrt{25}} = \frac{0.91}{5} = 0.18$$

In this case, the deviation from zero is:

$$\frac{0.3}{0.18} = 1.67 \text{ S.E.}$$

This is less than 2 S.E. and is not even significant at the five per cent level.

The student will see that little reliance can be placed upon results from small samples.

In previous discussions, it has been emphasized that the sample size must be reasonably large. We next consider a test which can be applied to samples of all sizes.

The χ^2 test of significance

Its purpose is to find whether the results obtained from a sample are likely ones, on the basis of some theory or law. The sample results will be termed *actual* ones, while the results suggested by our theory are termed *expected* ones. If the actual results are identical with those expected, there is no problem. Such correspondence does not prove the theory, as was pointed out in Chapter 14, but there is no difference to explain away.

If, however, there is a difference between fact and theory, the question arises as to whether the differences could be due to sampling fluctuations, or whether they are too large to be explained in this way, in which case they are *real* or *significant*.

In other words, we need an indication of whether the sample results fit the theory. For this reason, such tests are often called *goodness of fit* tests.

The χ^2 test (pronounced *kigh*-squared) is so-called because it is represented by the small Greek letter *chi*. From its wide variety of uses, we select two which commonly appear in examination questions.

APPLICATION TO FREQUENCY DISTRIBUTIONS

Example 2:

Table 17.1 gives the experimental results obtained from throwing a dice, together with those predicted by theory. In this instance, it is the theory of probability. Since a dice has six sides, we should expect that the chance of getting any particular result would be one in six, assuming the dice to be unbiased. That is, the frequencies of occurrence of each number (1–6) should be the same, and equal to $600/6 = 100$.

The value of χ^2 in any particular case is given by:

$$\chi^2 = \Sigma \frac{(F_A - F_E)^2}{F_E} \qquad \text{(F.17.2)}$$

where F_A = the actual frequency in each case,

$\qquad F_E$ = the theoretical or expected frequency in each case,

and $\qquad \Sigma$ = 'the sum of . . .', as before.

In other words, the difference between actual and theoretical in each class or group is squared, and the result divided by the expected frequency for that group. These individual results are then totalled for all the groups.

Table 17.1 Calculation of χ^2 for 600 throws of a dice

1	2	3	4	5
Score	*Actual frequency*	*Expected frequency*	$F_A - F_E$	$\dfrac{(F_A - F_E)^2}{F_E}$
1	80	100	−20	4
2	90	100	−10	1
3	100	100	0	0
4	105	100	5	0.25
5	110	100	10	1
6	115	100	15	2.25
TOTALS	600	600	0	8.5

Note 1. The total of column 4 gives a check upon the accuracy of the differences, since these should always total zero.

Note 2. Because the differences are squared, the result of column 5 is always positive. In the case where, for each class, there is complete agreement between fact and theory, all the differences will be zero, and the value of χ^2 is zero.

Note 3. The total of column 5 gives the value of χ^2 which, in this case, = 8.5.

Interpretation of the result We have seen that, where the actual and expected frequencies coincide, the value of χ^2 will be zero. In all other cases, it will be some positive amount, which increases with the difference between fact and theory.

We must now ask ourselves whether any particular value of χ^2 could have arisen by pure chance—i.e. could it be due to sampling fluctuations?

To answer this question, we make use of special tables, as we did in the case of the normal curve. But before we can understand these tables, we must introduce a new idea, namely, *degrees of freedom* (written D/F).

DEGREES OF FREEDOM

The significance of any particular value of χ^2 depends upon the number of degrees of freedom used in its calculation. To understand what this means, let us refer to column 2 of Table 17.1, i.e. the frequencies. Suppose we were to fill in this column by pure chance, by putting down any number, say, which sprang to mind. Bearing in mind that the total must be exactly 600, we have perfect freedom regarding the frequencies we insert in the first five lines, but the final frequency (score of six) is then taken out of our hands, since this will have to be the number which ensures that our total is exactly 600.

In other words, we have *five degrees of freedom* in this particular table. In general, for frequency distributions of this kind, the number of degrees of freedom is *one less* than the total number of classes.

The χ^2 table gives values of χ^2 arranged in rows, and each row corresponds to a particular number of degrees of freedom. The columns of the table are each headed by a figure of probability (P). These correspond to the previous notions of Chapter 14, i.e. $P = 0.1$ means a probability of 1 in 10, and so on (see Appendix 2). For example, referring to Table 17.1, we saw that there were five degrees of freedom. The χ^2 table, for 5 D/F, shows that, for $P = 0.1$, the value of $\chi^2 = 9.24$.

This means that a value of χ^2 as big as, or bigger than, 9.24 could arise from sampling fluctuations. The calculated value for Table 17.1 was 8.5, so the deviation from theory is very likely to be due to sampling fluctuations; i.e. there is no reason to suppose that the dice used was in any way biased.

Example 3:

Application to contingency tables

Sometimes, the subject of investigation is the association, or correspondence, between two or more attributes. Information of this kind is usually arranged in a table (called a *contingency table*) as below:

Table 17.2 Sample of adults, showing smoking habits by sex

	Smokers	Non-smokers	Total
Males	A 160	A 40	200
	E 120	E 80	
Females	A 440	A 360	800
	E 480	E 320	
TOTAL	600	400	1000

In Table 17.2, the A's (*actuals*) are the sample results. The marginal totals give a cross-check on the arithmetic, and add up to the grand total of 1000—i.e. the sample size.

The problem to be decided is whether the figures suggest any association between the two attributes of smoking (or non-smoking) and sex (male or female).

Step 1 We first put forward a theory, or hypothesis, to explain the results, and calculate the theoretical or expected frequencies (on this assumption) for each of the four subdivisions of the table.

In all such cases, we adopt the null hypothesis, as in *Example 1*; i.e. we assume that there is *no association* between sex and smoking. If this is so, we would expect the same proportion of males among the smokers as among the non-smokers since, according to our theory, *being* a male has no influence upon smoking. The proportion of males in the sample is 200/1000 and applying this to the total number of smokers, gives us an *expected number* (E) of male smokers of 200/1000 × 600 = 120. This figure has been entered in the table to explain the next step. (It is not necessary to do this in practice.)

If 120 males are smokers, then the rest are non-smokers. We can therefore place (200 − 120) = 80 in the next space.

Considering next the female smokers, we see that out of a total of 600 smokers, we have 120 males. Therefore, the rest (600 − 120) = 480 are female smokers. This only leaves the female non-smokers, and these must amount to 320, so as to give us our total of 800 females.

We have now filled in all our expected frequencies, in such a way that the marginal totals are still in agreement.

Step 2 We must now calculate the value of χ^2, as follows:

Table 17.3

Class	Actual frequency (A)	Expected frequency (E)	A − E	$\dfrac{(A - E)^2}{E}$
Male smokers	160	120	40	$13\frac{1}{3}$
Male non-smokers	40	80	−40	20
Female smokers	440	480	−40	$3\frac{1}{3}$
Female non-smokers	360	320	40	5
TOTALS	1000	1000	0	$41\frac{2}{3}(\chi^2)$

Step 3 We next determine the number of degrees of freedom in our calculations.

In our calculation of theoretical frequencies, we found the frequency for male smokers only. The remaining frequencies followed automatically from the various marginal totals. This suggests that there is only one degree of freedom, which is the case.

In general, for a contingency table for P rows and Q columns, the number of degrees of freedom is given by:

$$D/F = (P - 1)(Q - 1) \qquad \text{(F. 17.3)}$$

In the present case, there are only two rows and two columns (ignoring the marginal totals). Hence:

$$D/F = (2 - 1)(2 - 1) = 1 \times 1 = 1$$

Step 4 We now consult our table of χ^2 (Appendix 2). For 1 D/F we find that $P_{0.001} = 10.83$; i.e. the chances of getting a value of χ^2 as great as, or greater than, 10.83, through sampling fluctuations, are only 1 in 1000. This is very unlikely, and since we actually got a value of 41.67, such a result is virtually impossible.

Hence, we reject our hypothesis that there was no association between smoking habits and sex. The result is significant, and suggests that there *is*, in fact, such an association.

This is essentially a question of testing a manufactured product for *variability*, on the basis of successive samples, taken from the production line at regular intervals throughout the working day. **Quality control charts**

If the results are recorded on a chart, the inspector is provided with a cumulative picture of what is happening, and any tendency for the process to 'drift' can be easily seen. Furthermore, the chart provides:

1. An early warning system of any source of variation.
2. An aid to finding out whether a change is real, or only apparent.

The type of chart used will, to some extent, depend upon whether the aspect under control is an attribute or a variable (see Chapters 14 and 16).

CONTROL OF ATTRIBUTES

This method applies where it is possible to lay down certain standards, on the basis of which we can accept or reject a particular article. The rejected items are called *defectives* and the aim is to make sure that the *percentage or proportion of defectives* in the bulk of the production, remains within acceptable limits. The method used is essentially the application of sampling theory to binomial and Poisson distributions, as previously explained. The new features are the use of charts and the use of small samples, which are much more variable.

The expected percentage defective in the bulk production may be determined in advance, as a matter of policy, or it may be based on a series of tests. Suppose it is five per cent. If we take periodic samples of 20 items, we should therefore expect on average, one defective per sample. In fact, we might get as many as six defectives per sample, owing to fluctuations of sampling. We therefore lay down two limits for the fluctuation in the number of defectives found per sample.

One is called a *warning limit* or an *inner limit*, and is such that it will be exceeded, by chance, only 25 times in a thousand. This is really the five per cent level of significance which was discussed in Chapter 14, but here we are only concerned with *one tail* of the sampling distribution (two-and-a-half per cent).

The other limit is called the *upper limit* or the *action limit*, and the chances of exceeding this, through sampling fluctuations, are only 1 in 1000 (the 0.1 per cent level). If a sample result falls outside the first limit, it is an 'amber light', and further samples must be taken as soon as possible to confirm that it was simply a chance result. If the action limit is exceeded, the production is stopped, because this result is evidence that the production process has changed significantly.

The number of defectives for our two limits could be calculated on the basis of our formulae F. 14.1–14.3 and F. 14.6–14.8, according to whether the binomial or the Poisson type of distribution was considered more appropriate. In practice, however, tables are available which enable us to turn up the result immediately. The best known are those issued by the British Standards Institution, which lays down a uniform procedure.

Figure 17.1 shows a typical quality control chart of this type, with the results of 20 tests filled in. The last sample shows clearly that something has changed, and production would be stopped.

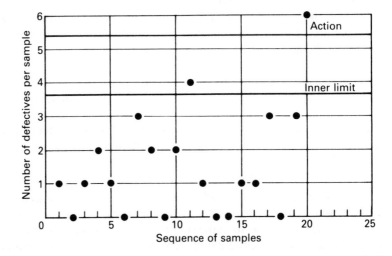

Fig. 17.1 Control chart for defectives; samples of 20. Expected percentage is five (one per sample)

CONTROL OF VARIABLES

This method is used when the quality under control is a continuous variable, e.g. a dimension or a weight. Mass-production, and the necessity for spare parts to be interchangeable, make it essential for all dimensions to lie within certain limits.

The two statistical measures used for such purposes are the average (usually the arithmetic mean) and the spread of dispersion (either the range or the standard deviation). The appropriate theoretical distribution is now the normal curve, since we are concerned with a continuous variable rather than one which is discrete.

(a) Control chart for averages

The inner control limits are fixed at A.M. ± 1.96 S.D./$\sqrt{n}$, i.e. we find the standard error of the arithmetic mean and take 1.96 times this figure as the five per cent limit of sample fluctuations.

The action limit is A.M. ± 3 S.D./$\sqrt{n}$ or 3 S.E.

Note. In this case we have *two* inner and *two* outer limits, since a component which is too small in some dimension is just as much a reject as one which is too large to fit (Fig. 17.2).

$$\text{Process A.M.} = 0.5 \text{ cm. S.D.} = 0.01 \text{ cm. Samples of 16}$$

$$\text{Inner limits} \quad 0.5 \pm \frac{1.96(0.01)}{\sqrt{16}} = 0.5 \pm 0.0049$$

$$\text{Outer limits} \quad 0.5 \pm \frac{3(0.01)}{\sqrt{16}} = 0.5 \pm 0.0075$$

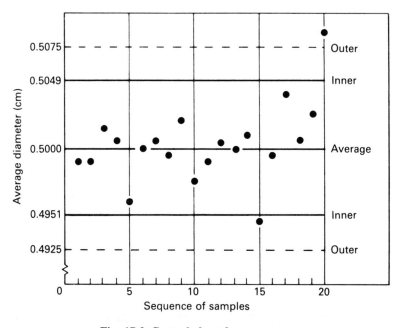

Fig. 17.2 Control chart for averages

(b) Control charts for standard deviation

Although the mean of a sample may be within the inner control limits, the spread of the individual items within a sample may be excessive. Some control of dispersion is therefore needed, and the best method is to use the standard deviation.

An estimate of the standard deviation of the universe is first made, using the results from a succession of samples. The standard deviations of individual samples, which are the values to be plotted on our chart, will fluctuate according to the normal curve, and tables are again available to give the inner and outer control limits.

As long as sample values remain within the limits, the process is under control. The moment we get a value outside the action limit, the process is stopped (see Fig. 17.3).

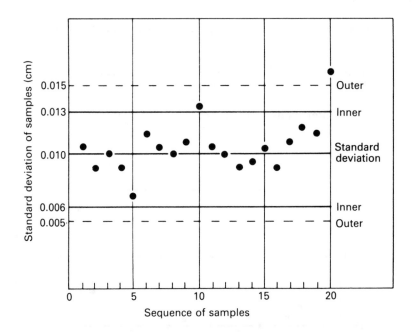

Fig. 17.3 Control chart for standard deviation

(c) Control charts for the range

An alternative method for controlling dispersion is to use the range as a measure of spread. This is more easily calculated, as the samples are taken, and is more easily understood.

The control limits for the range can be found directly from the standard deviation of the universe, using suitable conversion tables, or they can be found by calculating the average range of a group of samples, and again referring to appropriate tables for the limiting values.

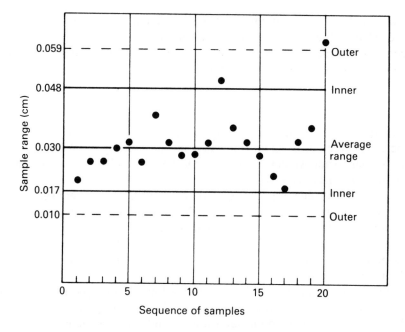

Fig. 17.4 Control chart for range of samples

Fig. 17.4 is a typical chart for the range.

Process S.D. = 0.01 cm. Samples of 16

For calculation of limits we use S.E. of S.D. $\left(\text{i.e. } S.E._{S.D.} = \dfrac{S.D.}{\sqrt{2}} \right)$

Thus, outer limit = 0.01 cm $\pm \dfrac{3(0.01)}{\sqrt{2} \times 16}$ = 0.01 cm $\pm$ 0.005 cm

Note. Limits are not symmetrical.

Process S.D. = 0.01. Process range = 0.03. Samples 16

Use and interpretation

We mentioned that this measure is of limited value. In addition to a large sample size (100 or more pairs), the value of r must be small—certainly less than 0.5. **Standard error of r**

When these two conditions are satisfied, the sampling distribution of r is roughly normal, and our various levels of significance can be used.

265

When r is large, the sampling distribution departs from the normal and is skewed. Hence, the above measure is not used.

There are alternative tests of significance which are used in such cases, but they are outside the scope of this book.

χ^2 test

The student may wonder why the table of χ^2 is limited to just a few levels of probability. This is because the shape of the χ^2 curve varies according to the number of degrees of freedom (indicated in the table by v). When $v = 1$, the curve is J-shaped and when $v = 30$, it approaches normality, but for intermediate values the curve has marked positive skewness.

Hence, for each value of v from 1 to 30, the curve has a different shape, and to tabulate fully the various degrees of probability would require 30 separate tables on the lines of Appendix 1 of the normal curve.

To avoid such a complication, the χ^2 table gives, for each value of v, the values of χ^2 corresponding to a few selected significance levels (values of P), which are sufficient in practice.

Strictly speaking, the idea of degrees of freedom is also applicable to standard deviation, and when calculating this measure, the divisor should be $(n - 1)$—the number of degrees of freedom—and not n—the total frequencies. When the total of frequencies is a larger number, the difference is of no importance, but for small samples, the divisor of $(n - 1)$ is used.

Contingency tables

Students usually have no difficulty in calculating χ^2 but the interpretation of the result often troubles them. Bearing in mind that we always use the null hypothesis, i.e. that χ^2 should really be zero, it follows that if P (probability) is small, then the divergence from zero is unlikely to be due to sampling fluctuations; i.e. we must reject the hypothesis. But since this is an assumption of *no association* between the attributes, we are driven to the conclusion that *association exists*, and that the result is significant. Hence, a value of P as small as 0.05 or smaller is always significant. A very large value of P is also significant ($P = 0.9$) since it means that there is little chance of obtaining a smaller value of χ^2. That is, the agreement between fact and theory is too good to be true.

Calculation of χ^2

This is not as difficult as it seems, since only one (any one) theoretical frequency need be found for a 2×2 table. The rest follow from the totals.

Furthermore, examination of Table 17.3 (*Step 2*) shows that the differences are all the same (in this case 40). The sign is immaterial, since the results are squared.

Therefore, we need not find the remaining theoretical frequencies at all. Having obtained one difference, we use this throughout.

It is even possible to calculate χ^2 without knowing this difference, if the student cares to memorize a simple formula. Consider the general case of a 2×2 table as follows:

Table 17.4

	A	Not A	Totals
B	a	b	c
Not B	d	e	f
TOTALS	g	h	n

It does not matter what A and (Not A) represent, as long as every item in our sample falls into one group or the other. For example, in Table 17.2, these columns were smokers and non-smokers. Provided there are only two columns, everything is one or the other. Similarly, provided there are only two rows, everything is B or (Not B). In Table 17.2, there were males and non-males (i.e. females).

The letters in the spaces represent the actual frequencies given in the question.

The value of χ^2 is now given by:

$$\chi^2 = \frac{n(ae - bd)^2}{c.f.g.h} \qquad (F.17.4)$$

Cross-multiply the frequencies, and take the smaller product from the larger (the reason for this follows later). Square the difference and multiply the result by the total frequencies (sample size). Divide the answer by all the marginal totals multiplied together.

Example 4:
Taking the actual frequencies of Table 17.2 we have:

Table 17.5

	A	Not A	Totals
B	160	40	200
Not B	440	360	800
TOTALS	600	400	1000

$$\chi^2 = \frac{1000 \times [(360 \times 160) - (40 \times 440)]^2}{200 \times 800 \times 600 \times 400}$$

$$= \frac{1000 \times [57\,600 - 17\,600]^2}{200 \times 800 \times 600 \times 400}$$

$$= \frac{1000 \times [40\,000]^2}{200 \times 800 \times 600 \times 400} = \frac{1000 \times 40\,000 \times 40\,000}{200 \times 800 \times 600 \times 400}$$

$$= \frac{1000 \times 16}{2 \times 8 \times 6 \times 4} = \frac{1000}{24} = 41.67 \text{ as before}$$

P × Q
contingency
When the number of rows or columns exceeds two (e.g. a 3 × 2), the formula cannot be used and the theoretical frequencies for each space in the table must be found as follows, using the null hypothesis, as on page 256.

Example 5:   **Table 17.6**

Factory	Quality grade of output			Totals
	1	2	3	
A	A 50	A 15	A 30	95
	E 44.3	E 15.8	E 34.9	
B	A 20	A 10	A 25	55
	E 25.7	E 9.2	E 20.1	
TOTALS	70	25	55	150

Assuming the raw material quality is identical, is there a significant difference in quality of production between the two factories? As in Table 17.2, the A's are *actual* results, and the E's are those *expected* on the basis of our theory.

Step 1 Since there are only 2 D/F [2 − 1)(3 − 1)] we need only calculate two of the theoretical frequencies (see *Example 3*, above).

Step 2 Taking space A.1, we multiply the marginal totals and divide by the sample size, i.e:

$$\frac{95 \times 70}{150} = 44.3$$

For space A.2, we have:

$$\frac{95 \times 25}{150} = 15.8$$

The rest of the spaces can then be filled in from the marginal totals.

Step 3 We then calculate each portion of χ^2 as in Table 17.3 giving us a total χ^2 of 3.9 (approximately).

Step 4 We consult our table of χ^2 as before. For 2 D/F we find that $P_{0.5} = 1.39$ and $P_{0.1} = 4.61$. We have *not* reached the five per cent level of significance ($\chi^2 = 5.99$) which is the minimum.

Hence we conclude the differences are due to sampling fluctuations.

The χ^2 distribution is a continuous one, like the normal curve, whereas the frequencies in a table such as Table 17.5 are discrete quantities. This is likely to introduce an error when the total (n) is small. A closer result is obtained in such cases by making a correction to the individual frequencies of the table. This consists of subtracting $\frac{1}{2}$ from those actual frequencies which exceed the corresponding expected frequency, and adding $\frac{1}{2}$ to those which fall short. The calculation then proceeds as before. **Yates's correction**

When using formula F. 17.4, this becomes:

$$\chi^2 = \frac{n\,(ae - bd - \frac{1}{2}n)^2}{c.f.g.h} \qquad \text{(F. 17.5)}$$

Note 1. Take the smaller product from the larger in the numerator, so that the correction gives a smaller answer.
Note 2. This correction is only needed for 1 D/F and for sample sizes under 100.

A simpler method of setting up control charts is to use a single control limit, based upon a probability of 0.005, i.e. the chances of a sample result falling outside it, due to sampling fluctuations, are 1 in 200. **Quality control**

This method is mainly intended for sampling of attributes where we are only concerned with the percentage or proportion of defectives.

It may be adapted for the control of variables, e.g. dimensions, where it is possible to test the product by using gauges. Two of these are used, one providing a maximum permitted dimension, and the other being set for the minimum dimension. All products which pass through the first are acceptable, in the sense that excessive dimensions are rejected. Those which *do not* pass through the second are equally acceptable, since they are above the minimum. Those which pass through it are rejected.

These devices are known as 'go' and 'no-go' gauges, and by their use the excessively variable items are rejects, so that the method of percentage rejects may still be used.

Further sampling with computer packages

We shall learn two ways of calculating χ^2 with MINITAB. In some circumstances we may already have 'reduced' our survey data, like the sex and smoking data of *Example 3* on page 259, to the form of a contingency table. Where this is the case we can read the contingency table into the MINITAB worksheet, and then use the CHISQUARE command: **MINITAB— quick CHI-SQUARE**

```
MTB > READ C1–C2
DATA > 160 40
DATA > 440 360
```

```
DATA > END
MTB > CHISQUARE C1–C2
```

Of course, PC MINITAB users can avoid READ and instead input the data in 'data edit' mode, accessed with ⟨Esc⟩. Notice that there are only four data items. These are the cell counts of Table 17.2. They represent the counts or frequencies of the four possibilities implied by the two dichotomous variables—male smokers, male non-smokers, female smokers, female non-smokers. MINITAB produces the contingency table and calculates χ^2.

MINITAB— useful TABLE

Our above reworking by MINITAB of *Example 3* is a little unrealistic since it assumes that we have already reduced the 1000 observations to a 2×2 contingency table. Why do this ourselves when the MINITAB package can do it quicker and better? The MINITAB command that produces contingency tables is TABLE. It is highly versatile and can be embellished with many subcommands, including the subcommand CHISQ.

In our survey dataset, we have recorded—in column C1—men as '1' and women as '2'. In column C2 we have recorded the smoking habits for each respondent; smokers are recorded as '1' and non-smokers as '0'. Each column is 1000 observations long. The two columns have been named 'SEX' and 'SMOKING', respectively. We can produce a simple contingency table from these 1000 observations on two variables with:

```
MTB > TABLE C1 C2
```

The χ^2 statistic is obtained by adding the CHISQ subcommand. Remember that in MINITAB subcommands are typed after adding a semicolon to the command line. We tell the package to execute the procedure by ending the (final) subcommand with a full stop:

```
MTB > TABLE C1 C2;
SUBC > CHISQ.
```

The package outputs the contingency table and adds:

```
CHI-SQUARE =   41.667   WITH D.F. = 1
```

More MINITAB tables

The TABLE command can be used simply to produce a 'one-way' frequency table. This is similar to FREQUENCIES in SPSS, e.g:

```
MTB > TABLE C1
```

Producing one-way, two-way and three-way tables is so important in most business statistics that it is worth dwelling a moment on other features of the TABLE command in MINITAB. To discover the full range of subcommands available, key HELP TABLE while in MINITAB. Employing the *Example 3* data (sex and smoking) again, here is a fuller use of TABLE with three subcommands:

```
MTB  > TABLE C1 C2;
SUBC > COUNT;
SUBC > ROWPERCENTS;
SUBC > TOTPERCENTS.
```

In this illustration each of the four cells of the contingency table contains three values:

1. The cell count or frequency (as in the first use above).
2. Each cell expressed as a percentage of the row total.
3. Each cell expressed as a percentage of the overall total of 1000 observations.

The TABLE command has other features when used with appropriate subcommands. TABLE can be used very elegantly to divide the sample into subgroups where the second variable of interest is a continuous one. Suppose that in *Example 3* we had also recorded weekly income (£s) for each of the 1000 respondents. Income is stored in column C3, and has been named 'PAY'. We use the subcommand MEAN with TABLE to calculate the average weekly income for the two sexes separately:

```
MTB  > TABLE C1;
SUBC > MEAN C3.
```

The output is as follows:

```
ROWS:  SEX

       PAY
       MEAN

    1  295.34
    2  187.50
  ALL  209.07
```

The SPSS[x] procedure to print contingency tables is CROSSTABS. With the appropriate STATISTICS keyword it will also calculate the χ^2 statistic. **SPSS[x] cross-tabulation**

In *Example 5* on page 268, we have two pieces of information on each of 150 units of production—the factory where they are made, and the quality of output. The data have been stored in an SPSS[x] system file as the variables 'FACTORY' (A = '1', B = '2') and 'QUALITY' (1,2,3). We wish to reproduce Table 17.6, plus the χ^2 statistic. The system file has the file name 'GRADES'. The following is a complete SPSS[x] job or run:

```
FILE HANDLE    INDATA / NAME = 'GRADES'
GET            FILE = INDATA
CROSSTABS      FACTORY BY QUALITY
OPTIONS        3 4 5
STATISTICS     1
FINISH
```

In this command or run file, the system file GRADES is called. The CROSSTABS procedure generates a two-way table. The OPTIONS keyword specifies options 3, 4, 5 (see SPSS^x manuals for a full list). These are the option numbers to print each cell (in addition to its count or frequency) as a row percentage, as a column percentage and as a percentage of the total of observations in the entire table. The STATISTICS keyword calls for a χ^2 calculation (see manuals for a full list of statistics available). The detail on the FILE HANDLE command that declares the name of the system file (GRADES) and its temporary 'handle' (INDATA) internal to the SPSS^x job, will vary between computer installations. See Chapter 12 for a revision of SPSS^x system files.

The output from the run appears in Fig. 17.5 below:

		QUALITY			
COUNT					
RDW PCT					ROW
COL PCT					TOTAL
FACTORY	TOT PCT	1	2	3	
	1	50	15	30	95
		52.6	15.8	31.6	63.3
		71.4	60.0	54.5	
		33.3	10.0	20.0	
	2	20	10	25	55
		36.4	18.2	45.5	36.7
		28.6	40.0	45.5	
		13.3	6.7	16.7	
COLUMN		70	25	55	160
TOTAL		46.7	16.7	36.7	100.0

CHI-SQUARE	D.F.	SIGNIFICANCE	MIN E.F.	CELLS WITH E.F. < 5
3.92407	2	0.1406	9.167	NONE

NUMBER OF MISSING OBSERVATIONS = 0

Fig. 17.5 Cross-tabulation of factory quality

Contingency tables are also produced in SPSS/PC+ with CROSSTABS. **SPSS/PC+**
Various subcommands are possible. A simple but useful form of SPSS/PC+ **cross-**
CROSSTABS for *Example 5* (factory and output quality) is: **tabulation**

```
CROSSTABS  FACTORY BY QUALITY / CELLS / STATISTICS = CHISQ.
```

After the CROSSTABS command, we specify the two variables to be used. If
the CELLS subcommand appears without further specification each cell of the
contingency table will contain four values. Each value expresses the cell in a
different way—as a count or frequency, as a row percentage, as a column
percentage and as a percentage of the total of observations for the entire table.
Finally, the STATISTICS subcommand specifies CHISQ, and ends the procedure with a period or full stop.

The MINITAB TABLE command and the SPSS CROSSTABS command **Complex**
may be used additionally to produce three-way and even higher-level tables. If **cross-**
we had a sex and smoking datafile that also held the variable 'ALCOHOL' on **tabulation**
the respondents' (alcohol) drinking habits (high, medium, low, non: 3, 2, 1, 0)
we could produce a three-way classification:

```
TABLE         'SEX' BY 'SMOKING' BY 'ALCOHOL'   (. . . in MINITAB)
    CROSSTABS  SEX BY SMOKING BY ALCOHOL  (. . . in SPSS)
```

These two commands each produce *four* two-way tables of SEX BY SMOK-
ING—one for each of the four categories of the *control* variable ALCOHOL (3,
2, 1, 0).

The use of TABLE with the MEAN subcommand in MINITAB allows us to **Comparing**
examine subgroups (e.g. by sex) of a continuous variable like income or **means**
expenditure (see above). Comparing means between groups in this way is
achieved in SPSS^x with the BREAKDOWN command, and in SPSS/PC+ with
the MEANS command. The general format of these commands is similar to that
of CROSSTABS:

```
BREAKDOWN PAY BY SEX                (. . . in SPSS^x)
    MEANS     PAY BY SEX.              (. . . in SPSS/PC+)
```

Statistical significance tests are available with both versions.
Like CROSSTABS, both BREAKDOWN and MEANS can specify further
control variables, e.g:

```
BREAKDOWN PAY BY SEX BY REGION
```

An elegant and simple way of achieving this double control in MINITAB is to create from the two control variables a two-way contingency table in which the cell contents are the mean values of the 'PAY' variable:

```
MTB > TABLE 'SEX' BY 'REGION';
SUBC > MEAN 'PAY'.
```

With five categories of the variable 'REGION' and two of 'SEX' there are 10 cells in the table. The TABLE command here outputs 10 subgroup means—one for each possible combination of SEX and REGION.

Further graphical applications

Method

In Chapters 6 and 7, we studied the basic principles of pictorial and graphical representation. We now consider some specialized applications of these ideas, which are popular in examinations.

There are two common applications of this device.

THE GANTT CHART

Alternative titles for the Gantt chart are *machine loading chart* or *progress chart*.

This type consists of two horizontal bar charts for each period of time (usually a day or a week). One, which is simply a heavy line, indicates the *planned* production or running time, and alongside this a bar chart represents the *actual* figures. Any discrepancy reveals a loss of production and is investigated. The chart may also show a *cumulative* comparison for the week as a whole.

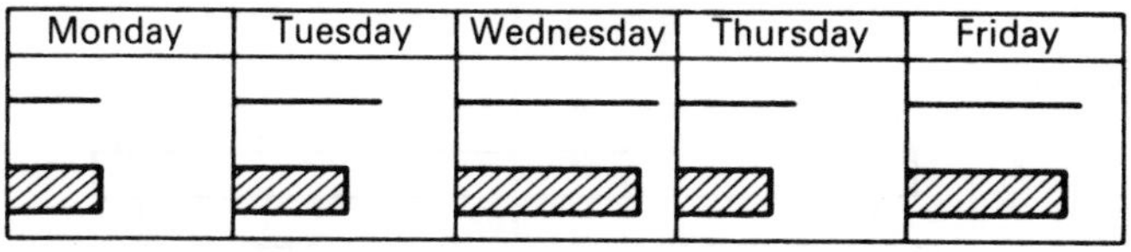

Monday	Tuesday	Wednesday	Thursday	Friday

Fig. 18.1 Gantt chart showing production record

Figure 18.1 shows the sort of result which may be obtained. The horizontal scale may be in percentage form, in which case the planned production for Monday was 40 per cent of capacity. This is shown as a thin line. The actual production fell short of target on every day except Monday, and the reasons for this would be investigated.

In Fig. 18.2 we have cumulative figures for a week, subdivided for each machine. Here again the horizontal scale could be running hours, or percentage of possible time. This type is useful for allocating rush orders which might come in. For example, there is ample surplus capacity on Machines 2 and 4 which could be utilized. The code letters show the reason for the failure to achieve the planned running time.

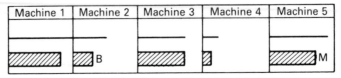

Code letter B = Breakdown; M = Material shortage

Fig. 18.2 Cumulative weekly total of machine loading

THE POPULATION PYRAMID

This type of bar chart is illustrated in Fig. 18.3. It consists of a series of horizontal bars (one for each age group), the length of each one being proportional to the number of people in that age group. There are really two separate diagrams, one for males and one for females, which are placed together for each comparison. If the number of births remained fairly constant over a long period (say $2\frac{1}{2}$ millions for each sex), one would expect each successive bar to be shorter than the one below it, since there would be a progressive reduction in numbers due to death; i.e. some of the $2\frac{1}{2}$ million would never reach the second bar (5–10 years). In this sense, one could say that the complete diagram, for both sexes, should resemble a stepped pyramid.

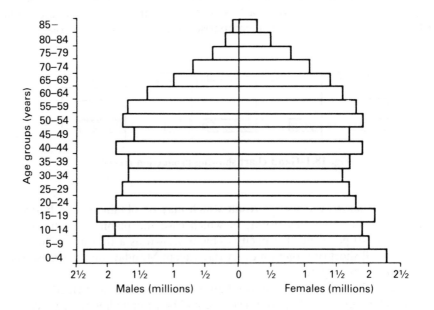

Fig. 18.3 Population pyramid showing age distribution of UK population

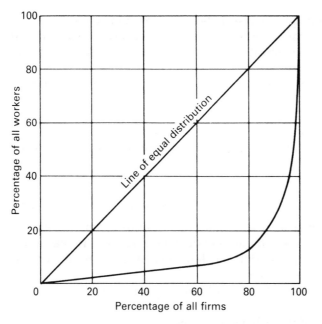

Fig. 18.4 Lorenz curve showing distribution of workers in industry

The two best known applications are the *Lorenz curve* and the *Z* (or *Zee*) *chart*.   **Cumulative curves**

THE LORENZ CURVE

This is a graph to show the interrelation of two cumulative percentages. That is, it involves a comparison of two variables, which are usually presented in the form of frequency distributions.

Example 1:

Table 18.1 Distribution of workers in manufacturing industry by size of firm

1	2	3	4		5	
			Percentages		*Cumulative percentages*	
Number of workers	*Number of firms (thousands)*	*Number of workers (ten thousands)*	*Number of firms*	*Number of workers*	*Number of firms*	*Number of workers*
Up to 10	150	75	73	9	73	9
11–24	15	26	7	3	80	12
25–49	15	51	7	6	87	18
50–99	12	81	6	9	93	27
100–249	9	135	4	16	97	43
250–499	3	114	2	14	99	57
500–999	2	105	0.5	12	99.5	69
1000–	1	261	0.5	31	100	100
TOTALS	207	848	100.0	100		

The two frequency distributions are given in columns 1–3. In column 4, we express the frequencies as percentages of their respective totals. It is permissible to round off the individual percentages, so as to make them total exactly 100 per cent in each case.

In column 5, a cumulative total is found for each of the percentage columns.

The information is then plotted on a graph, in which each axis is of equal length and is scaled in percentages (see Fig. 18.4).

The line of equal distribution (sometimes called the 45° line) is then drawn in, by joining the origin to the intersection of the 100 per cent lines.

THE Z CHART

This is a particular type of time chart or historigram. The intervals of time may be days, weeks or months, and the total extent of the chart is usually one year, although it is permissible to show several years' results on one diagram, for purposes of comparison.

It is really a combination of three separate diagrams:

1. A graph of the individual figures.
2. A cumulative total, or ogive, of the figures for the year (or each individual year).
3. A moving annual total (M.A.T.), starting with the annual total of the previous year, and adjusting this figure for each period, as the individual figures are available. The M.A.T. is the same idea as the trend figure (see Chapter 12), except that we do not divide by the number of items to obtain a moving average.

The scale used for 2. and 3. is usually smaller than that used for 1. since the magnitude of the figures being plotted is different, and it is better to adopt a more open scale for the original data so as to bring out more clearly the individual movements.

The scales to use will depend upon the intervals of time. For monthly figures, the scale for 2. and 3. could be five times that of the actual data, while for weekly figures it might be 20 times.

The chart resembles the letter Z, and hence its name.

Example 2 (see Table 18.2):

Note 1. The total sales for Year 1 (January–December) are 288 (£ thousands). The moving annual total from February Year 1 to January Year 2 (12 months) is £288 – January Year 1 + January Year 2 = £288 – 28 + 22 = £282. Subsequent totals are obtained by discarding the Year 1 figure and adding the corresponding Year 2 figure.

Hence, for February, we have £282 – £24 + £28 = £286; for March, £286 – £20 + £30 = £296, and so on.

Table 18.2 Sales figures of a firm (£ thousands)

Details	Jan.	Feb.	Mar.	Apr.	May	June	July	Aug.	Sept.	Oct.	Nov.	Dec.
Sales Year 1	28	24	20	18	22	23	20	19	24	28	30	32
Cumulative sales	28	52	72	90	112	135	155	174	198	226	256	288
M.A.T.	—	—	—	—	—	—	—	—	—	—	—	288
Sales Year 2	22	28	30	22	26	28	30	26	30	24	32	38
Cumulative sales	22	50	80	102	128	156	186	212	242	266	298	336
M.A.T.	282	286	296	300	304	309	319	326	332	328	330	336

Note 2.   There is a check on the arithmetic in the December column, since the cumulative sales for Year 2 should equal the moving annual total for December of that year.

The graph of the three series is shown in Fig. 18.5.

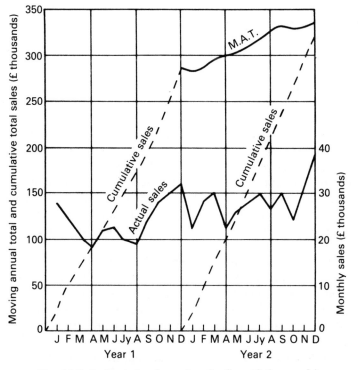

Fig. 18.5 Z chart showing sales of a firm (£ thousands)

Use and interpretation

The Gantt chart

This is mainly used for the control of machines and of production processes. Ideally, there would be one chart per machine or process, and each chart would run for a week, the bars or blocks being drawn in each day, at the end of the run.

Apart from their value in indicating discrepancies between plan and performance, they have many other uses. They give a pictorial survey of the whole range of manufacture, and the extent to which there is spare capacity. If they are filed away over a period, they also provide a source of reference on comparative reliability of different types of machine, and on the efficiency of material flow and production planning.

The population pyramid

This gives an indication of the age distribution of the population at any particular time, and is valuable for the planning of social and economic projects.

Reference to Fig 18.3 indicates clearly the so-called *population bulge* where certain bars are actually larger than those beneath them, and which is often referred to in the press. In fact, there are two bulges—one around 40–50 years, which forecasts an increasing proportion of pensioners in the near future, and the other in the 20–25 years group.

This effect is brought about whenever there is a 'population explosion'—i.e. a significant increase in the birth-rate over a few years.

The bulge rises up the pyramid and, in time, works its way out at the top. Actually, the term 'pyramid' is seldom correct for a typical population structure, because females survive longer than males, and consequently the two halves of the pyramid are rarely symmetrical.

The Lorenz curve

This was originally suggested as a method of measuring distribution of wealth, but it gives a useful picture of divergence from the average. The more the curve departs from the line of equal distribution, the greater the concentration of divergent items at that point. For example, Fig. 18.4 reveals that some 90 per cent of the firms (i.e. the small ones) employ collectively only 20 per cent of the labour force. Alternatively, one could say that 80 per cent of the workers are employed by only 10 per cent of the firms (the big ones). Were the distribution uniform, one would expect, say, half the number of firms to employ half the labour force. The curve is, perhaps, less useful than a coefficient of dispersion, since it does not provide a numerical measure of the divergence.

The Z chart

If applied to sales, as is usually the case, the Z chart shows at a glance how sales are progressing month by month and reveals any seasonal fluctuations. The cumulative curve shows how the performance to data compares with the target figure, and with the corresponding point of time in previous years. The moving annual total gives an annual indication, with seasonal fluctuations removed, and is a picture of the trend.

Applying this analysis to Fig. 18.5, we can see, from actual sales, that March and April are poor months, while November and December seem to be above average. The cumulative total for Year 2 was ahead of the previous year from March onwards, and overall trend is, on balance, upwards.

More graphs by computer

The MINITAB package has 11 different commands that produce quality control charts of the type introduced in Chapter 17. Control charts for attributes and charts for variables may be produced.

MINITAB control charts

Suppose that we wish to use MINITAB to reproduce the control chart displayed in the previous chapter as Fig. 17.2 (Control chart for averages). We have data on the diameter of test components. There are 20 samples of these components, with 16 units in each sample—i.e. we have observations on a total of 320 components. The measurement data are stored in the MINITAB worksheet in column C1 ('DIAMTR'). Column C2 ('SAMPLE') records which sample the test component belongs to (numbered 1 to 20 in blocks of 16). The first few rows of our worksheet look as follows:

DIAMTR	SAMPLE
0.4987	1
0.5011	1
0.5056	1
"	"
"	"

The sequence of commands to reproduce Fig. 17.2 is:

```
MTB  > XBARCHART C1, C2;
SUBC > SLIMITS 1.96, 3.
```

The XBARCHART command calls a control chart for averages from the 320 rows of data stored as column C1. Column C2 is cited as the location of the sample identifier. The command ends with a semicolon to indicate that a subcommand will follow. The SLIMITS subcommand specifies where the upper control limits ('UCL') and lower control limits ('LCL') will be calculated and drawn. Here, just as in Fig. 17.2, we have two sets of UCL and LCL, the inner set at ± 1.96 S.E. and the outer set at ± 3 S.E. The subcommand sequence ends with a period or full stop. Labels, like those in the original figure, might have been supplied with further subcommands of XLABEL, YLABEL, TITLE and FOOTNOTE, each of which is followed by 'text' (within single quotation marks).

Other sorts of control chart are available in MINITAB. The standard deviation control chart of Fig. 17.3 in the previous chapter can be reproduced with SCHART to replace XBARCHART. A control chart for range is produced with the RCHART command (see Fig. 17.4).

Control charts for attributes are available with commands like NPCHART (binomial) and UCHART (Poisson). Refer to the MINITAB manuals for details of which charts are available, and for the specific and the general subcommands in this MINITAB suite of 11 statistical process control procedures.

Spreadsheet graphics

Most spreadsheets allow the user to produce graphs and charts from the data in the spreadsheet. In the example below we use the AS-EASY-AS package to produce a bar chart that replicates the female side of the population pyramid displayed in Fig. 18.3 on page 276.

Having called up AS-EASY-AS, we enter the data into the worksheet. In column A we enter the age group labels (0−, 5−, etc.). Each label is preceded by a single quotation mark to indicate to the package that the digits are entered as text rather than as quantitative data for use in calculations. In column B we enter the number of women falling into each of the 18 age groups. We therefore fill two columns with 18 rows:

```
. . . . . . A / . . . . . . B / . . . . . . C / . . .   .   .
1      0−        2.28
2      5−        2.01
3      .
4      .
```

After data entry we press ⟨/⟩ to call the pop-up menu of AS-EASY-AS facilities. With the cursor keys we move the highlight to 'Graphics' and press ⟨**RETURN**⟩. From the Graphics sub-menu that now replaces the main menu we use the cursor keys to highlight 'Type' and again confirm with ⟨**RETURN**⟩. A sub-sub-menu springs up with a list of the types of graphs and charts available. From this list we highlight 'Bar' (for bar chart) and confirm with ⟨**RETURN**⟩. We have now selected the type of representation that we want.

The next step is to define the ranges of the variables that are to constitute the chart. The Graphics sub-menu from which we previously chose 'Type' is once more displayed on the screen. We move the highlight to 'X' and press ⟨**RETURN**⟩. This starts the sequence in which we define the range that is to be the 'independent' or 'X' variable. Once the sub-menu has cleared from the screen we move the cursor to highlight cell 'A1' as the 'active cell' (i.e. the first row of the column of age labels). When A1 is highlighted, we key a period or full stop, ⟨.⟩, to fix the start of the X-range. We then use the 'cursor down' arrow key to cell A18, the cell of the last age label. As we do this, the whole block of cells A1–A18 becomes highlighted. When the highlight reaches A18 we press ⟨**RETURN**⟩ to confirm that this is the last cell of the X-range.

Our next step, when the Graphics sub-menu appears, is to define the range of the 'dependent' or 'A' variable in one worksheet. We highlight and select 'B' from the list and proceed to define B1–B18 as the range of the dependent variable, just as we did in the previous paragraph for the X variable.

Once this second range has been defined, the Graphics sub-menu reappears. To use the chart that has been defined, use the cursor arrow keys to highlight

'View' and press ⟨**RETURN**⟩. The chart is drawn on the screen. To get back to the worksheet from the chart simply press ⟨**Esc**⟩. If we wish, we can re-enter the menu system with ⟨ / ⟩ in order to add labels and other embellishments to the chart. We select 'Graphics' again, and from subsequent sub-menus we choose 'Options' and then 'Titles'. From the Titles menu we can choose to add two main titles ('First', 'Second') and to label the independent (X) axis and the dependent (Y) axis.

The result of our work is the AS-EASY-AS bar chart shown below in Fig. 18.6. This reproduces the right-hand side of Fig. 18.3.

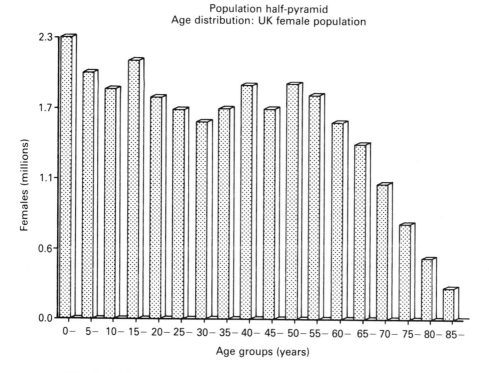

Fig. 18.6 AS-EASY-AS bar chart of population half-pyramid

The chart is printed on paper by choosing 'Plot' from the Graphics sub-menu. From the next sub-menu that appears we highlight and select (with ⟨**RETURN**⟩) 'Go'. This sends the chart to the printer. Note that the scale points in Fig. 18.6 are not those of Fig. 18.3. This is because the package has used automatic scaling in the AS-EASY-AS version. From the many menu options within the Graphics sub-menu we could have chosen our own scaling.

The spreadsheet offers other graph and chart types, including the scattergram ('X-Y' graph) and pie chart. If you get lost in the menu and sub-menu system simply keep pressing ⟨**Esc**⟩ to get back to the worksheet. Other spreadsheets offer graphics in a similar way to AS-EASY-AS.

19

Financial mathematics

Method

In this book, as far as we have dealt with the firm, we have been concerned mainly with statistical techniques of internal control and presentation. We must now look at some techniques which may help us in setting up a business or a project, e.g. discovering statistics that will help us to decide *whether* to invest.

This branch of calculation is known as 'investment appraisal.' It is a well-known tool of the cost accountant and it should be familiar to the economist who studies marginal costing. The techniques involved are simple and are well within the grasp of the statistician who may know little about accountancy or economics. The student will be thankful to learn that very little extra technical knowledge is required for the understanding of these techniques.

Arithmetic progressions
An arithmetic progression is a series of numbers where the difference between the numbers is the same.

Example 1:

$$3, 7, 11, 15, 19 \text{ (the difference is 4)}$$

$$4\tfrac{1}{2}, 3, 1\tfrac{1}{2}, 0, -1\tfrac{1}{2}, -3 \text{ (the difference is } -1\tfrac{1}{2})$$

The difference, as you can see, can be added or subtracted. The series of numbers can be expressed in whole numbers, fractions, decimals, etc.

In calculations of the simple interest on a sum of money invested we can use the arithmetic progression.

Example 2:
If £100 is invested for four years at a simple interest rate of five per cent per annum, at the end of four years, the total amount accumulated would be:

$$£100 + £5 + £5 + £5 + £5 = £120$$

i.e. the total amount would have grown, in an arithmetic progression, thus:

$$£100, £105, £110, £115, £120$$

If we call the total amount accumulated A, and the original investment P, and r the rate of interest, then:

$$P, (P + Pr), (P + 2Pr), (P + 3Pr), (P + 4Pr)$$

is the arithmetic progression.

The formula of the last term of this (when $t =$ time in years) is:

$$A = P\,(1 + tr), \text{ or } (P + tPr) \qquad \text{(F. 19.1)}$$

Thus, in Example 2:

$$A = £100 \left(1 + 4 \times \frac{5}{100}\right) = £120$$

Note 1. r is expressed as a fraction of 100, and t can be calculated as a fraction if the time involves half or quarter years, months, weeks or even days.

Note 2. The student is invited to compare formula F. 19.1 with the equation for a straight line ($Y = a + bX$) given in Chapter 11.

If we wished to work a problem in 'reverse' and wanted to find out how much we would have to invest (P) at simple interest (r) to provide ourselves with a certain target amount (A) at the end of a specified period (t years) the formula is simply:

$$P = \frac{A}{(1 + rt)} \qquad \text{(F. 19.2)}$$

In Example 2:

$$P = \frac{£120}{\left(1 + \dfrac{5}{100} \times 4\right)} = £100$$

The term P is usually called the 'present value' of A at a simple interest rate (r) in t years from now.

Geometric progressions

A geometric progression is a series of numbers where the difference between the numbers is found by *multiplying* the preceding number by a fixed amount (often called the 'common ratio').

Example 3:

> 4, 8, 16, 32, 64 (each number is multiplied by 2
> to get the following number)

> 160, 40, 10, $2\frac{1}{2}$, $\frac{5}{8}$ (each number is multiplied by $\frac{1}{4}$)

Geometric progressions can increase or decrease (as can arithmetic progressions) but they do so in *more* than a constant amount. (At this stage, the student might refer back to pages 116 to 117 where arithmetic and geometric series were used).

The geometric progression is used in calculations of compound interest. It is often the case, where a sum of money has been invested, for the interest payment (say, in a bank or building society account) to be reinvested, rather than to be withdrawn and spent. If this is so, then, in our previous example, we would begin with $P = £100$ in the first year, and then, at the beginning of the second year we would be investing a different P (= £105) and so on, with an increased P at the beginning of each year. Thus, our geometric series would be:

$$P, P(1 + r), P(1 + r)(1 + r), P(1 + r)(1 + r)(1 + r), \text{etc.}$$
$$(£100, (£105), (£110.25), (£115.7625), \text{etc.}$$

i.e. $P, P(1 + r), P(1 + r)^2, P(1 + r)^3, P(1 + r)^4$, etc.
The 'common ratio' is $(1 + r)$.

The formula we obtain from the last term is:

$$A = P(1 + r)^t \text{ (i.e. to the 'power' of } t) \qquad \text{(F. 19.3)}$$

Thus, in Example 2:

$$A = £100 \left(1 + \frac{5}{100}\right)^4 = £100(1.05)^4$$

Using logarithms this gives:

$$\log 100 + 4 \log (1.05)$$

$$\text{Therefore } A = £121.60$$

Comparing this with the return on a simple (arithmetic) interest we would clearly expect more if we reinvested our interest than if we did not. The extra return is £121.60 − £120 = £1.60.

The formula for compound interest can also be used for asking what sum we would have to invest (P) at compound interest (r) to get a target amount (A) at the end of a certain period (t years). The formula for this reverse process is:

$$P = \frac{A}{(1 + r)^t} \qquad \text{(F. 19.4)}$$

In our Example 2:

$$P = \frac{121.60}{\left(1 + \dfrac{5}{100}\right)^4} = \pounds 100$$

Once again, P is known as the 'present value' and the method of obtaining P (by our formula) is known as 'discounting' the amount A.

When a businessman or woman wishes to invest in a new project or a new product he or she will want to know how much to expect in returns and, therefore, in profits. The issue is not always as simple as this. Usually, when a business has spare funds for investment there will be more than one idea for a variety of new products, or new projects, and the choice has to be made as to which will be the most profitable. It is difficult to look into the future (as was explained in Chapter 12, Time series), but there are different techniques for doing so, some of which are better than others.

These techniques fall into two classifications:

Investment appraisal

1. Conventional methods
   (a) Pay-back method
   (b) Average rate of return (or accounting return based on initial investment) method
2. Discounting methods
   (c) Net present value method
   (d) Internal rate of return method

These are the four methods that we shall describe here, although there are variations. We shall use one example to illustrate the four different methods, and we shall use the concepts of the progression of arithmetic and geometric series.

CONVENTIONAL METHODS

(a) Pay-back method
This method is the most commonly used in investment appraisal. It is simple and crude, and it is defined as the time it takes for an investment to generate sufficient returns to pay back the original investment in full.

Example 4:
Supposing a firm has a choice of five projects, each with a different initial cash investment, and each project investment is calculated to bring a certain flow of cash by every year end, as follows:

Table 19.1

	End of year	A	B	C	D	E
				Projects		
Initial cash investment	0	110 000	100 000	210 000	180 000	50 000
Cash inflows	1	0	20 000	70 000	0	20 000
	2	20 000	20 000	70 000	0	20 000
	3	20 000	20 000	70 000	0	20 000
	4	20 000	20 000	70 000	0	20 000
	5	20 000	20 000	70 000	0	20 000
	6	20 000	20 000		0	
	7	20 000	20 000		0	
	8	20 000	20 000		0	
	9	20 000	20 000		0	
	10	20 000	20 000		900 000	
	11	20 000				
	12	20 000				
	13	20 000				
	14	20 000				
	15	20 000				

The method is simply to add the cash inflows to find out which sum would pay back the initial cash investment.

Note. In Project D, we do not receive *any* cash inflow until the end of year 10! The answers can be put down thus:

Table 19.2

Project	A	B	C	D	E
			Pay-back periods		
Years	6.5	5.0	3.0	10.0	2.5

Clearly, the best project *on this method* is Project E which takes the *shortest* time (2.5 years) to repay the initial cash investment.

(b) Average rate of return method (ARR)
This method simply averages the annual cash inflow, less a deduction for depreciation, over the initial cash investment, or, if investments are made over a number of years in a project, over the average investment.

$$\frac{\text{Average annual cash inflow} - \text{Average annual depreciation}}{\text{Initial cash investment}} \quad \text{(F. 19.5)}$$

For Project A, this would be (as a percentage):

$$100 \times \frac{\dfrac{£280\ 000}{15} - \dfrac{£110\ 000}{15}}{£110\ 000} = 10.30 \text{ per cent}$$

The student can easily work out the complete list as follows:

Table 19.3

Project	A	B	C	D	E
		Average rates of return			
Percentage	10.30	10.00	13.33	40.00	20.00

Obviously, Project D gives the *greatest* ARR, and is therefore to be preferred *on this method*.

DISCOUNTING METHODS

For these two methods we use what we have already learned of the progression of series. If you were offered, say, £100 now, or £100 in three years time, which would you take? Only a foolish person would turn down £100 now. The reason is quite obvious. You could, at simple interest, increase your £100 to £115 in three years time by investment, or at compound interest, to £115.7625, as we found earlier. (Also, in a time of inflation, £100 in three years time would be worth less anyhow!).

As we saw before, the present value of a future amount depends on two things:

1. The number of years it takes to receive it.
2. The rate of interest at which the original investment is made. (See F. 19.4).

Figure 19.1 shows, in graphical form, how the time and the interest rate affect the present value of £1. Reading from the graph we can learn, for example, that the present value of £1 to be received in 20 years time at 5 per cent is about 38p (a), or, in 10 years time at 20 per cent is about 16p (b). We can use this discounting technique to improve on the conventional methods just described, to give a more accurate appraisal of investment decisions.

Note. The *curves* of the geometric progressions on the graph would form a series of *straight* lines if these were arithmetic progressions (simple interest). The

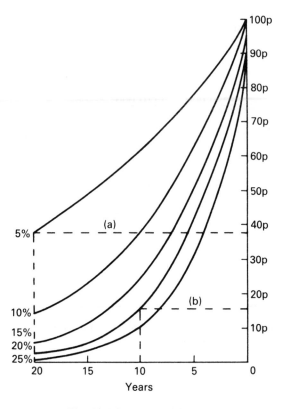

Fig. 19.1 Present value of £1

student is invited to calculate from formula F. 19.2 (100p = A) what P would be in 20 years time at 5 per cent and in 10 years time at 20 per cent, and to draw the straight lines on Fig 19.1.

(c) Net present value (NPV) method

In this method the cash inflows of the five projects set out in Table 19.1 are discounted to their present value, i.e. in Year 0 (the time of the initial cash investment), and the total discounted cash flows are subtracted from the value of the initial cash investment. The formula is:

Σ (annual cash inflow × discount factor) − initial cash investment   (F. 19.6)

There are two pieces of information we require before we can calculate this. First, we have to know the rate of interest that we must use. This is usually fixed at a rate which would give shareholders in the firm a rate at least equal to what they could expect from alternative investment opportunities outside the company. In our example we are going to fix this at 16 per cent. Second, we need a table which will give us 'discount factors' calculated for various interest rates for varying periods of years, so that we can discount the cash inflows to their present values. These are given in Appendix 3.

Note. The student is invited to check the measurements on Fig. 19.1 by the 'present value factors'. Hence:

20 per cent for 10 years = 0.1615 (16p approximately)
5 per cent for 20 years = 0.3769 (37½p approximately)

Let us start with Project C. The initial cash investment is £210 000 and the cash inflows are £70 000 for each of the first five years. In Appendix 3 (Present value factors), we look at the column headed 16 per cent to find the discount factors by which we must multiply, year by year, the five cash inflows of £70 000, thus:

Table 19.4

Year 1	£70 000 × 0.8621 = £60 347
Year 2	£70 000 × 0.7432 = £52 024
Year 3	£70 000 × 0.6407 = £44 849
Year 4	£70 000 × 0.5523 = £38 661
Year 5	£70 000 × 0.4761 = £33 327
TOTAL	£229 208

These are totalled, and the initial cash investment is then subtracted from this total.
Note. The initial cash investment may be *larger* than the total of the discounted cash inflows, in which case, the result will be *negative*.

£229 208 − £210 000 = £19 208

This result means that the firm could borrow, for example, £229 208 for Project C, use £210 000 to begin the project (i.e. by investing it in the project) distribute £19 208 to its shareholders, and, by the end of the project (end of Year 5) the loan would have been paid off. The project is acceptable as it allows *some* profit (£19 208) to the shareholders—but is it to be preferred to the other four projects? We must work out, in the same manner, the other four projects. This would clearly be laborious work. Fortunately, we have another table to help us in Appendix 4 (Cumulative present value factors).

If we look at this table, under the column headed 16 per cent, and along row Year 5, we shall find the figure 3.274. As the cash inflows for each year in Project C are the same (£70 000) we can simply multiply £70 000 by 3.274. This gives us £229 180.
Note. This figure is a little lower than the figure obtained previously (£229 208) but this is merely because the 'cumulative' table's figures have been rounded off (If you want to prove this, add up the first five years' figures under the column 16 per cent in Appendix 3).

In the following table, the calculation for all five projects have been made, using the cumulative table of present value factors:

Table 19.5

Project A	(£20 000 × 5.575 − 0.862) − £110 000	= −£15 740
Project B	(£20 000 × 4.833) − £100 000	= −£3340
Project C	(£70 000 × 3.274) − £210 000	= +£19 180
Project D	(£900 000 × 0.227) − £180 000	= +£24 300
Project E	(£20 000 × 3.274) − £50 000	= +£15 480

Clearly Project D has the highest profit (+£24 300) and is therefore to be preferred.

Note. Obviously projects with negative profits are not worth considering.

There are some points which the student should note about these calculations:

1. All the cash inflows in our example are assumed to be received at the end of Year 1. The cash outflow (i.e. the initial capital investment) occurs at the beginning of Year 1, i.e. at the 'end of Year 0'. The discount factor for the initial cash investment is therefore 1000 (because the present value of £1 in Year 0 is £1).
2. In Project D, as only one cash inflow is received (i.e. in Year 10) the present value factor of 16 per cent for Year 10 is used, and we cannot use the cumulative table because there are no years previous to Year 10 to take into account.
3. In Project A, if we use the cumulative table for £20 000, at 16 per cent, for 15 years (5.575) we must subtract the present value factor for Year 1 (0.862) because no cash inflow was received for that year.

(d) Internal rate of return (IRR) method

This method of discounting is really more common in business than the NPV method. Calculation in this method can be stated as follows: that rate of interest which, when used to discount the cash flows of a proposed investment, reduces the present value of the project to zero. (It is sometimes known as the *discounted yield* method). In this method we do not require to assume our interest rate of 16 per cent.

We can therefore put the IRR formula in the form of an equation:

$$\text{annual cash inflow} \times \text{discount factor} = \text{initial cash investment}$$

(i.e. making the terms 'equal' is the same thing as cancelling out one side in terms of the other).

If we rearrange this formula slightly to get:

$$\frac{\text{initial cash investment}}{\text{annual cash inflow}} = \text{discount factor} \qquad \text{(F. 19.7)}$$

we can substitute our known quantities in the left-hand side and find the discount factor. Then, proceeding along the *row* of our cumulative table for the number of years for which cash has been flowing in, we can find the discount factor we have calculated and proceed to the top of that particular column to find the interest rate.

Example 5:
For Project A, substitution would give us:

$$\frac{£110\ 000}{£20\ 000} = 5.5 \text{ (discount factor)}$$

Looking up 5.5 in the cumulative table for the number of years £20 000 has been flowing in (i.e. 15 years − 1 year = 14 years) we find the figure *nearest* to 5.5 to be 5.577 (i.e. 6.462 (15 year row) *minus* 0.885 (1 year row)). The interest rate at the head of this particular column is 13 per cent. This is the internal rate of return for this project that will reduce the initial cash investment (approximately) to zero. The student will probably realise that this method is the reverse process of the NPV method.

The table of IRR calculations for all the projects is given below:

Table 19.6

Project A	$\dfrac{£110\ 000}{£20\ 000}$	= 5.5 (for 14 years)	IRR = 13 per cent approx.
Project B	$\dfrac{£100\ 000}{£20\ 000}$	= 5.0 (for 10 years)	IRR = 15 per cent approx
Project C	$\dfrac{£210\ 000}{£70\ 000}$	= 3.0 (for 5 years)	IRR = 20 per cent approx
Project D	$\dfrac{£180\ 000}{£900\ 000}$	= 0.2 (for the 10th year on the present value table	IRR = 17 per cent approx
Project E	$\dfrac{£50\ 000}{£20\ 000}$	= 2.5 (for 5 years)	IRR = 29 per cent approx

Note. Obviously Project A requires the lowest rate of return to make any profit, therefore this will be the favoured project *on this method*.

Use and interpretation

Arithmetic progressions

The main use of the arithmetic progression is in the calculation of simple interest in business transactions. Its use is limited, however, and the student should assume, unless 'simple interest' is stated in the calculation, that the compound interest calculation is required. The reason for this is that it is more realistic to assume that interest will be reinvested in an efficiently run business or fund.

Geometric progressions

Geometric progressions are used in the calculation of compound interest in many business affairs. These range from the calculation of interest on bills of exchange, securities and annuity funds, tax arrears, bank and building society interest, etc. They will probably be familiar to the student and therefore we shall give only one example of the application of series, i.e. depreciation.

It should be clear to the student by now that compound interest produces a higher rate of growth in the total amount for the same interest rate on the same principal. Furthermore, the produce of compound interest will become proportionately greater:

1. if the length of time money is invested increases.
2. the more frequently the interest is reinvested, i.e. if we reinvest (recalculate) the interest, say, monthly instead of yearly, or weekly instead of monthly.

Depreciation

This is the gradual reduction in the value of capital assets in a firm (e.g. plant, machinery, vehicles, etc.) and it depends on the passage of time, the rate at which assets become obsolete, and the wear and tear to which they are subjected. Accountants place different emphases on these three factors and this gives rise to a variety of methods. We shall deal only with three of these methods:

THE STRAIGHT LINE METHOD

This is a simple averaging method which allocates an equal cost of an asset to each period (e.g. year) during which it is used. For example:

$$\text{Cost of machine} = \text{£}1000$$

$$\text{Period of use} = 20 \text{ years}$$

$$\text{Annual depreciation} = \frac{\text{£}1000}{20} = \text{£}50$$

Accountants who use this method assume that the machine gives equal value in *use* every year of its life. The method gives an *arithmetic* series of 'written off' values.

THE REDUCING BALANCE METHOD

In this method, the asset is 'written down' to its economic *value* at the end of each successive year of its life (i.e. it is assessed at its market value—which, when it is no longer usable as a machine, may be its scrap value). For example:

$$\text{Cost of machine} = \text{£}1000$$

$$\text{Period of use} = 20 \text{ years}$$

First year of depreciation = £1000 − £80 = £920

Second year of depreciation = £920 − £90 = £830 etc.

The method *could* give a *geometric* rate of 'written down' values, e.g. the reduction of each successive year's *reduced value* by a fixed percentage or fraction.

Note. These are only two of the simpler methods of assessing depreciation. The student should bear in mind that in today's changing world, a machine could become obsolete whilst it is still relatively new, therefore its value may suddenly drop sharply. On the other hand, if the inflation rate is very high, a machine could actually be worth *more* when it is older than a new one of the same type.

SINKING FUND METHOD

1. If a firm wishes to build reserves (i.e. savings) to invest in plant or machinery in the future, it may, like a private individual, save a regular sum each year. Naturally it would 'invest' it in a safe security while the fund builds up, and in its accounts the money would be entered in a special 'reserve' or sinking fund account. Say, £100 was saved each year for four years at five per cent, then, using formula F. 19.3, each £100 would build up like this:

$$£100 \text{ (invested at beginning of Year 1)} = £100 \left(1 + \frac{5}{100}\right)^4 = £121.6$$

$$£100 \text{ (invested at beginning of Year 2)} = £100 \left(1 + \frac{5}{100}\right)^3 = £115.8$$

$$£100 \text{ (invested at beginning of Year 3)} = £100 \left(1 + \frac{5}{100}\right)^2 = £110.5$$

$$£100 \text{ (invested at beginning of Year 4)} = £100 \left(1 + \frac{5}{100}\right)^1 = £105$$

Therefore, at the beginning of Year 5 we should have the *addition* of these four separate 'streams' of investments, i.e. £452.9. Therefore, our sinking fund would have produced £452.9.

2. A more useful kind of sinking fund is one where we wish to produce a *certain* figure, e.g. to replace plant which costs a given amount of money, or to provide a depreciation fund. Here we have a *target* to aim for in a *given* number of years. How much must we invest annually?

    Say we wish to produce £1000 (using the same rate of interest—five per cent) at the end of four years instead of our £452.9. Then, as our unknown *annual* investment sums would grow at the same rate over the same period of time, their growth would be proportional to £100:£452.9, i.e. x (our unknown annual investment): £1000.

We can say $\dfrac{100}{452.9} = \dfrac{x}{1000}$. In other words, x will be equal to $\dfrac{1}{4.529} \times$ £1000 = £220.9.

Therefore, our contribution to the sinking fund, in order to raise £1000 at five per cent in four years time, would have to be £220.9 per year. This is an example of a sinking (or depreciation) fund which uses a discounting method.

Note 1. The student may compare formula F. 19.3 with the calculation in 1. above and formula F. 19.4 with the calculation in 2. It will be seen that the figures have been discounted on precisely the same basis, but simply in reverse order.

Note 2. The above example is a simple one as we wish only to illustrate the concept. The labour of calculation in more complex examples is greatly lightened by the use of a book of tables such as, Yeats, A., White, J. and Skipworth, G., *Financial Tables* (Stanley Thornes (Publishers) Ltd, Cheltenham, 1978), which contains a large variety of tables to calculate sinking funds, compound interest, annuities, etc.

Investment appraisal

The methods described above are ones which do not look at production costs per unit in the manner of the average and marginal costs of the economist. What we are concerned with are simply cash inflows (positive) and cash outflows (negative). If we wish to make a table of these, a very simplified one might look something like this:

Table 19.7

Cash outflows (negative)	Cash inflows (positive)
Building and land costs	Sales receipts
Machinery costs	Government investment grants
Raw material costs	Tax allowances
Corporation tax	Scrap value of assets
Labour costs	

Note 1. *Depreciation* does not involve any actual cash flowing in any direction, therefore if we have recorded the initial investment as a negative cash flow, and the money we have received for the scrap value (or residual value) as a positive cash flow, then we have *assumed* depreciation by the difference between these figures.

Note 2. In Example 4, comparing Projects A to E, we describe a simple situation, uncomplicated by tax and government grants, etc., but clearly, in a 'real life' problem, these would be entered by an accountant—and appropriately discounted to the time they were paid or received.

Note 3. We could have a situation where we are comparing projects that all have common costs, e.g. if all projects have the *same amounts* of common costs (such as all using the same building) then we could ignore these.

Our simple example has, however, given the basis of the techniques used, and the student will have observed that the four methods gave three different results:

Table 19.8

Method	Preferred project
Pay-back	E
Average rate of return	D
Net present value	D
Internal rate of return	A

In order to understand this apparently puzzling result, let us look at some of the advantages and disadvantages of each method in turn.

PAY-BACK METHOD

Advantages

1. Easy to understand.
2. Simple to operate.
3. Most useful in industries in which rapid technological change takes place since it gives preference to projects which have the shortest pay-back period.
4. Useful to businesses which need to have a high liquidity (i.e. have a high need for ready cash or assets which they can quickly turn into ready cash) for frequent reinvestment or for crises.

Disadvantages

1. The 'discounting' factor is ignored, i.e. the factor that money received now is worth more than money received later.
   Note. Some accountants *do* use discounting factors in the pay-back method. If they do, it makes the pay-back period longer, of course.
2. It fails to take note of long-term profitability, i.e. it misses completely the earnings (cash inflows) arising *after* the end of the pay-back period (some highly profitable investments do *not* always pay high returns in the early years).

AVERAGE RATE OF RETURN METHOD

Advantages

1. Easy to understand.
2. Simple to operate.
3. Unlike the pay-back method, this method does attempt to measure *profitability*.

Disadvantages

1. The 'discounting' factor is ignored
2. As arithmetic mean averaging is used, no weight is given for fluctuations (e.g. long-term build-up to late high profits).

NET PRESENT VALUE METHOD

Advantages
1. The 'discounting' factor is recognized as part of the calculation.
2. Can be adjusted to take care of investment grants, taxes, inflation, different degrees of risk.
3. Depreciation is automatically allowed for.

Disadvantages
1. Appears to be more complicated than the two conventional methods, but is really quite easy to understand, and the labour of calculation is greatly eased by the use of prepared tables.
2. It is difficult to set discount rates for risk. It is even more difficult to set different discount rates for different projects with different degrees of risk.

INTERNAL RATE OF RETURN METHOD

Advantages
This method has, generally, the advantage of the NPV method.

Disadvantages
1. Rather more complex to understand and calculate.
2. When cash flows are irregular, requires extended calculation.

Conclusion   All appraisal methods suffer from our inability to predict the future, e.g. forecasting cash inflows. The discounting methods are obviously superior to conventional methods, particularly as they assess money's worth on a more scientific and reasonable basis. There are many other problems which weaken some of these methods, e.g. assessing the degree of risk, defining exactly what is the cost of capital, etc.

In our example, different results for different methods were obtained. To interpret these surprising results, the student must note four points:

1. To say that Project X is 'preferred' does not mean that *all* the others are rejected. It is, for example, possible that a firm in our example may have enough money to invest in *all* the projects except, perhaps, in Projects B and C (which none of the methods puts first).
2. It is best to rely on discounting methods, except where high liquidity in a fast changing industry is desired.
3. The job of the statistician and the accountant is to provide data (with all their faults and reservations) for decision-makers to act on, and to explain, if necessary, the bases on which different kinds of measures are compiled.
4. Finally, the student must realise that many investments are made not on the quantitative results of statistical or accounting measurements. Many investments are often made *despite* an *adverse* report from a quantitative appraisal. They may be made on environmental grounds, for industrial relations reasons, on grounds of legality or social responsibility, or an effect on public relations, quite apart from someone's 'hunch'.

Financial mathematics and spreadsheets

Spreadsheets are designed to handle financial mathematics easily. To illustrate this we have entered the *Example 4* data in Table 19.1 into the AS-EASY-AS spreadsheet package. To get to the stage shown by Fig. 19.2 below we have simply called up the AS-EASY-AS software and keyed the information into the worksheet of cells arranged in rows and columns. We have moved from cell to cell of the worksheet with the familiar cursor 'arrow' keys (e.g. $\langle \downarrow \rangle, \langle \rightarrow \rangle$) on the keyboard. (See the computing sections of Chapter 9 for a practical introduction to spreadsheets.)

Calculations using a spreadsheet

F1:Help 2:Edit 3:Macro 4:Abs 3 READY! 3 5:Goto 6:Window 7:Wp 9:Calc F10:Graph

C22: @AVG(C5..C19)

[A]	A/ Endyear	B/ Project A	C/ Project B	D/ Project C	E/ Project D	F/ Project E	G/ H
1	Endyear	Project A	Project B	Project C	Project D	Project E	
2 Invest	0	110000	100000	210000	180000	50000	
3							
4		−110000	−100000	−210000	−180000	−50000	
5 CashIn	1	0	20000	70000	0	20000	
6	2	20000	20000	70000	0	20000	
7	3	20000	20000	70000	0	20000	
8	4	20000	20000	70000	0	20000	
9	5	20000	20000	70000	0	20000	
10	6	20000	20000		0		
11	7	20000	20000		0		
12	8	20000	20000		0		
13	9	20000	20000		0		
14	10	20000	20000		900000		
15	11	20000					
16	12	20000					
17	13	20000					
18	14	20000					
19	15	20000					
20							
21							
22 AvCashIn		18666.66	20000	70000	90000	20000	
23 AvDeprec		7333.333	10000	42000	18000	10000	
24 ARR		10.30303	10	13.33333	40	20	
25							
26							
27 IRRFunc		0.132399	0.150984	0.198577	0.174618	0.286492	

Fig. 19.2 Data from Table 19.1 entered into a spreadsheet

Notice that we have entered the investment data twice—once as a positive figure in row 2 (just as in the original Table 19.1), once as a negative figure in row 4. This double entry is not strictly necessary, but is useful for the work below.

The two calculations that we have reproduced using the spreadsheet are the *average rate of return*, and the *internal rate of return*.

AVERAGE RATE OF RETURN

Let us first take the average rate of return (ARR) calculation, and focus on worksheet column C, the column that contains the data for Project A. We have decided to enter our own formulae for the ARR calculation. For Project A this work is undertaken in the cells C22, C23, C24. However, because we are using a spreadsheet, when we look at the worksheet on the VDU screen we see in these three cells only the results of our calculations:

$$18666.66$$
$$7333.333$$
$$10.30303$$

The formulae or functions that we actually keyed into these cells (C22, C23, C24) in order to produce instant answers in the worksheet, are hidden from view. However, they are not lost. They can be revealed by moving the cursor to column C, and then pressing the function key ⟨**F7**⟩ on the keyboard. You can see from the guideline at the top of the worksheet that the ⟨**F7**⟩ function key on the keyboard evokes the command 'Wp' (wordprocessor) in the spreadsheet. You can think of the 'Wp' as meaning to 'wipe away' the results from the worksheet to reveal the combination of data, labels and formulae that were originally keyed into the cells. In this case pressing the ⟨**F7**⟩ 'Wp' function key reveals in C22, C23, C24 the following:

$$@AVG(C5..C19)$$
$$+C2/15$$
$$100\star(C22-C23)/C2$$

When we printed the worksheet the cursor was resting on cell C22. That is why towards the top left-hand corner of the screen display we can see 'C22: @AVG(C5..C19)'. Placing the cursor on any cell immediately reveals its original keyed content in this way.

The sequence of three expressions reproduces the formula given as F.19.5 above. However, in this case we have to use the grammar of the command language understood by the spreadsheet. We keyed into C22 one of the package's built-in functions. These built-in functions all start with the '@' symbol. The @AVG function gives the arithmetic mean of all the data in the specified range. Here the range specified is columns C5 to C19 inclusive—i.e. we obtain the average annual cash inflow over the 15-year life of Project A. Into C23 we keyed an expression, '+C2/15', to give us the average annual depreciation of the initial investment (the £110 000 recorded in cell C2, without the £ sign, divided by 15). Finally, in C24 we keyed the expression to give us the average rate of return expressed as a percentage:

$$100 \times \frac{\text{average annual cash inflow } - \text{ average annual depreciation}}{\text{initial cash investment}}$$

In spreadsheet terms this has become:

$$100 \times \frac{C22 - C23}{C2}$$

and in the command notation of the spreadsheet, cell C24 contains:

$$100 \star (C22 - C23) / C2$$

Compare our spreadsheet results with those from the earlier working following Table 19.1. Our average rate of return results are the same—and so they should be!

For the second calculation, the internal rate of return (IRR), we have chosen to use the function @IRR, already built in to AS-EASY-AS, that will give us the internal rate of return from a single formula. The @IRR function returns the rate of return per period associated with a cash flow. The exact form of the function is:

$$@IRR(\text{Interest Rate, Range})$$

Here, 'range' means the range of cells that are to figure in the calculation. Despite the way that we earlier calculated IRR from formula F.19.7, the AS-EASY-AS function requires us to input an assumed interest rate (here 16 per cent or 0.16). In cell C27 of the spreadsheet, to obtain the IRR result for Project A, we enter '@IRR(0.16,C4..C19)' (without quotation marks). Cells C4 to C19 (inclusive) contain the cash-flow data for Project A. The cash inflows are given as the positive numbers in C5 to C19. Cell C4 contains the initial investment, here appropriately expressed as a *negative* cash-flow element.

Compare the IRR results given in the cells of row 27 in Fig. 19.2 with the workings of *Example 5* on page 293. Converting the spreadsheet results into percentage form (i.e. × 100), we get the same results for Projects A to E as before—an IRR of 13 per cent, 15 per cent, 20 per cent, 17 per cent and 29 per cent, respectively. We get the same results and come to the same conclusion—i.e. using the IRR method, Project A is preferred because it requires the lowest rate of return in order to make a profit.

To exit from AS-EASY-AS we press the front slash ⟨ / ⟩ key. This calls a pop-up menu. Using the down cursor ⟨ ↓ ⟩ key we move the highlight bar to 'File' and press ⟨**RETURN**⟩ or ⟨**ENTER**⟩. This, in turn, engages us in a short dialogue in which we save our spreadsheet to a file stored on diskette or hard disk. This done, we press ⟨ / ⟩ again, but this time take the highlight bar down to 'Exit'. Pressing ⟨**RETURN**⟩ takes us out of the AS−EASY-AS package via another short dialogue.

20

Vital statistics

Method

The early importance of vital statistics has already been noted in Chapter 1; indeed, they are the basic statistics of life in an economy, e.g. births, marriages and deaths. Many techniques already described in this book are commonly used by the demographer—the name given to the specialist in this branch of statistics. Sources for these figures are included in Chapter 21.

The student will hardly need to be reminded how 'vital' this group of statistics is to the politician, the sociologist or the economist, and of how essential such information is as a tool for planning. The student must be warned, however, that demography is an extremely sophisticated study and that an understanding of the basic methods used, and even the additional mathematical training essential to the actuary in life assurance, does not qualify anyone as more than an amateur in this field.

Briefly, we can say that living populations in any country depend on births, deaths, immigration and emigration. These four factors depend further on other factors. For example, the birth-rate depends on the sex ratio, the proportion of women aged between 15 and 45 years (the normal child-bearing period), the expected size of family, the rate of illegitimate births, etc. In addition, the death-rate may also affect the birth-rate; for example, if a war reduces the number of adult males in a country. Similarly, immigration and emigration may affect the birth-rate according to the age and sex of the kind of people who enter and leave a country. Other sets of factors may affect the death-rate and the emigration/immigration rates. The student might also think of social, political and economic factors which affect these rates. Do rising wages affect the desire for more children? Does membership of the European Community lead to greater migration? Does a 'permissive' society lead to a larger illegitimacy rate?

In line with the pattern of this book, we now examine some of the methods of statistics which are used in the field of vital statistics to analyse changes in basic population factors.

Population
302

Chapter 1 stressed the importance of exact definition. The term 'population' is defined in a variety of ways in official statistics, the principal ones being:

1. *De facto or home population* This means everyone in this country at the time. Therefore, it includes seamen in British ports, foreign and colonial forces here, and so on.
2. *Total population* This is the home population plus HM Forces overseas, less foreign forces in the UK.

CRUDE BIRTH- AND DEATH-RATES

These are simply the total (live) births or deaths during 12 months per 1000 of the population (home).

We can express them as follows:

$$\text{Crude birth-rate} = \frac{\text{total live births} \times 1000}{\text{total population}}$$

$$\text{Crude death-rate} = \frac{\text{total deaths} \times 1000}{\text{total population}}$$

Similarly we can use simple rates or ratios to express other vital relationships, and these are often given as percentages, or per 1000 of the appropriate population.

SEX RATIO

$$\text{Sex ratio} = \frac{\text{total number of females} \times 100}{\text{total number of males}}$$

MARRIAGE RATE

This can be expressed in a variety of ways:

1. Persons married per 1000 population (all ages).
2. Males married per 1000 unmarried females (15+).
3. Females married per 1000 unmarried females (15+), and so on.

ILLEGITIMACY RATE

Here again, we find rates for:

1. Illegitimate births (live) per 1000 unmarried women (15–44).
2. Illegitimate births (live) per 1000 live births.

INFANTILE MORTALITY RATE

$$\text{Infantile mortality rate} = \frac{\text{total deaths of babies up to one year old} \times 1000}{\text{total live births}}$$

AGE-SPECIFIC RATES

These are rates (births or deaths) for a specified age or age group (usually five-year) as opposed to crude rates which are based on totals (i.e. all ages).

The rates given above are, however, too simple to give more than a general indication of the forces which affect population in a country. It would be misleading, for example, to compare on a time series basis, England in 1820 with England in 1990 in the matter of the crude death-rate. Social and economic factors apart, the age structures of the two populations were different and a 'young' population would, by itself, tend to produce a lower death-rate than the 'ageing' population of today. Similar considerations would apply to the birth-rate.

When we come to compare different regions or towns within a country, and also when we seek to compare different countries, the results we would get might be totally misleading. Havarest, a seaside retirement town, and Youthville, a bustling new industrial town, might produce a high death-rate and low birth-rate, and a low death-rate and high birth-rate, respectively. How would Havarest and Youthville compare in their illegitimacy rates and in their sex ratios? It would be foolish to conclude that Havarest was less healthy than Youthville on the evidence of their crude birth- and death-rates.

In other words, any crude rate is the weighted arithmetic means of a series of age-specific rates, the weights being the actual numbers of people in each age group at the time. These weights will obviously differ for particular areas, or for the same area at different times.

**Standar-
dized
birth- and
death-rates**

The main purpose of statistics is *comparison*, and we can only compare like with like. This means that the weights used in our calculations must be identical, so that any differences in age/sex structure are removed.

The tables below give hypothetical details for our two towns of Havarest and Youthville.

Note. Column 8 always contains arbitrary 'model' rates chosen by the statistician himself.

Table 20.1 Death rates in two towns

1	2	3	4	5	6	7	8	9	10
Age group	Havarest			Youthville			Standard population (thousands)	Number of expected deaths in standard population if subject to mortality of Columns 4 and 7	
	Popu-lation (thous-ands)	Number of deaths	Death rate per 1000	Popu-lation (thous-sands)	Number of deaths	Death rate per 1000		Havarest (4) × (8)	Youthville (7) × (8)
0–10	10	30	3	20	100	5	17	51	85
10–20	10	10	1	18	54	3	13	13	39
20–40	20	40	2	40	120	3	30	60	90
40–60	30	420	14	15	225	15	27	378	405
60–80	25	875	35	6	240	40	12	420	480
80–100	5	650	130	1	150	150	1	130	150
TOTALS	100	2 025		100	889		100	1 052	1 249

Note 1. The age-specific death-rates are given in columns 4 and 7.
Note 2. The crude death-rates are 20.25 per 1000 for Havarest, and 8.89 for Youthville; that is:

$$\frac{\text{total column 3}}{\text{total column 2}} \text{ and } \frac{\text{total column 6}}{\text{total column 5}}$$

From these it would appear that Havarest is a more unhealthy place.
Note 3. The standardized death-rates are given by:

$$\frac{\text{total column 9}}{\text{total column 8}} \text{ and } \frac{\text{total column 10}}{\text{total column 8}}$$

It is clear from these that Havarest (10.52 per 1000) is actually healthier than Youthville (12.49 per 1000). The difference in the crude rates being due to the more youthful population of Youthville.

Life tables

The data collected during the Census of Population forms the basis for certain detailed studies by the Registrar General, one of which is the *life table*. These tables are produced in detail every 10 years, but *abridged* tables are prepared between censuses and appear in the *Annual Abstract of Statistics*, as well as the publications of the Office of Population Censuses.

The life table provides two basic facts. First, it traces the mortality experience of a theoretical population from birth to death.

This is done by stating (1) the probability that a person aged X will die within one year, and (2) the number who would survive to an exact age X, out of the original 100 000 (or 10 000) if they were subject, throughout their lives, to the death probabilities of (1).

Table 20.2 Extract from life table

Age, X (years)	Males		Females	
	Number surviving to age X out of 10 000 born	Average future lifetime for a person aged X (years)	Number surviving to age X out of 10 000 born	Average future lifetime for a person aged X (years)
0	10 000	68.5	10 000	74.7
10	9 735	60.4	9 795	66.5
30	9 580	41.2	9 719	46.7
50	9 057	22.8	9 353	28.0
70	5 561	9.4	7 341	12.3
80	2 340	5.4	4 495	6.8

Table 20.2 is an extract from an abridged table. This is based on a theoretical population of 10 000, but the *full* tables work on a total of 100 000.

From this we see that, from a hypothetical 10 000 males born, 9735 will survive to their 10th birthday (97.35 per cent) and 5561 will survive to their 70th birthday.

This assumes that the hypothetical population is subjected to the same mortality experience throughout as the *actual* population upon which the table is based.

In the second place, the life table enables a further set of data, called the *expectation of life*, to be calculated. This is the average future duration of life, for a person of specified age, if subjected to the death-rates of the table.

For example, quoting from the same table, a male aged 10 years has an average expectation of a further 60.4 years, and one aged 70 years has an expectation of a further 9.4 years.

It is important to realize that expectation of life always refers to some specified starting age.

The reader will note, from Table 20.2, that a male aged 0 years has an average life span of 68.5 years. It might seem from this that men reaching 50 years have only 18.5 years to live, on average. It can be seen from the table that they can expect a further 22.8 years. The explanation lies in the fact that a newly born baby has to face all the health hazards of a lifetime, and many will fall by the wayside. A man aged 50 years has already survived the risks of childhood and adolescence, and such persons have fewer remaining hurdles to surmount.

A simple example will make this clear. Suppose our population to consist of only five men, who die at ages of 50, 55, 59, 67 and 84 years. Their average life span from birth is $\frac{1}{5}(50 + 55 + 59 + 67 + 84) = 63$ years. This is their expectation of life.

Starting from age 60 years, we only consider those who are still alive, i.e. those who die at 67 and 84, and these live for a further 7 and 24 years respectively, i.e. an average of 15.5 years beyond 60. The expectation of life at age 60 is thus 15.5 years. In a full life table, the survivors are given for each sex, separately, and for each year of life, from birth to 105 years. The expectation of life is also given for each single year of age.

Reproduction rates

A problem of major importance is to forecast future population trends; birth-rates by themselves are a poor indication. The key factor is clearly the number of married women in the population who are between 15 and 44 (the reproductive years). The number of live births in a year, per 1000 such women, is the legitimate *fertility rate* but this will vary according to the age composition of the women, since their fertility varies according to their position in the age group 15–44 years.

The procedure adopted, in a simplified form, is to take (say) 1000 women and assume that they pass through the various age groups of the reproductive years. In so doing, they are assumed to be subject to current fertility rates, and will therefore produce a certain number of *female* babies for each age group (column 2 of Table 20.3.

Table 20.3 Gross and net reproduction rates

1	2	3	4
Age group	Number of female babies born to 1000 women in each age group	Number of survivors per 1000 female babies	Number of female babies reproduced by 1000 female babies
15–19	100	950	95
20–24	200	900	180
25–29	250	800	200
30–34	250	700	175
35–39	150	600	90
40–44	50	500	25
	1000		765

The total number of babies so produced, per 1000 women, is the *gross reproduction rate* of 1.000, i.e. column 2, divided by 1000.

Note. This calculation assumes that all the 1000 women will survive to age 44, and that all the female babies they produce will themselves survive to reach child-bearing age, and thereafter until age 44. This, of course, is not so.

Column 3 of Table 20.3 shows the number of survivors (to various age limits) of 1000 female babies.

Column 4 shows the number of female babies left behind by the original 1000 females, allowing for deaths on the way; e.g. out of 1000 female babies, 900 survive to age 20–24. As 1000 females pass through this group, they bear 200 female babies (column 2). *Hence*, 900 women will bear $\dfrac{900}{1000} \times 200 = 180$ female babies (column 4), and so on.

The total of column 4, divided by 1000, gives the *net reproduction rate* of 0.765. See the section on Use and interpretation for an explanation.

Comparisons of death- or birth-rates between different regions are often misleading, as we have seen, because of the different age and sex structures.

The construction of standardized rates, previously illustrated for deaths, is one method of overcoming this difficulty, but it assumes that we know the number of deaths in each region, classified by age. For many regional populations, this information is not always available and it is not possible to calculate local age-specific rates which can be applied to a standard population, as in Table 20.1. However, the age-specific rates (births or deaths) for England and Wales as a whole are known, and can be applied to local populations. The rates employed are usually those of the latest available census.

Area compara-bility factors

As applied to deaths, the procedure is as follows:

1. Take the numbers per age group in the region and multiply by the standard death-rate for England and Wales (for that particular age group).
2. Divide the total products by the total population (total weights). This gives the *expected* death-rate per 1000 for the region.
3. The standard death-rate of England and Wales is next found. (That is, the products of the age-specific rates and the actual numbers in each group of the standard population are totalled and divided by the total standard population.) This is given at the foot of column 4, Table 20.4.
4. The standard death-rate divided by the expected death-rate gives the *area comparability factor* (A.C.F.).

The area comparability factor is a standardizing factor, which, when multiplied by the crude death-rates for each region, will give a standardized rate for that region. Any two such rates for different regions or areas can then be compared, as they are based upon the same age structure.

Table 20.4 illustrates this method for our imaginary towns of Havarest and Youthville.

Table 20.4 Standardized death-rates in two towns

1	2	3	4	5	6
Age group	Population (thousands)		Standard death rates per 1000 for England and Wales (approximate)	Expected death on basis of death rates	
	Havarest	Youthville		Havarest (4) × (2)	Youthville (4) × (3)
0–9	10	20	3.4	34	68
10–19	10	18	1.0	10	18
20–39	20	40	2.3	46	92
40–59	30	15	14.9	447	223.5
60–79	25	6	39.2	980	235.2
80–99	5	1	137.0	685	137.0
TOTALS	100	100	12.6	2202	773.7

Note 1. The expected death-rates per 1000 are:

$$\text{Havarest} = \frac{2202}{100} = 22.02$$

$$\text{Youthville} = \frac{773.7}{100} = 7.74$$

Note 2. The standard death-rate for England and Wales is 12.6 per 1000. (See step 3. above for the derivation of this figure.)

Note 3. The area comparability factors are:

$$\text{Havarest} = \frac{12.6}{22.02} \simeq 0.57$$

$$\text{Youthville} = \frac{12.6}{7.74} \simeq 1.63$$

Note 4. The A.C.F. is a *standardizing* factor, which, when multiplied by the crude death-rates for each town, yields the standardized rates for the two towns. These are:

$$\text{Havarest} = 0.57 \times 20.25 \text{ (Table 20.1)} = 11.54$$

$$\text{Youthville} = 1.63 \times 8.89 = 14.5$$

These are published in the census reports on occupational mortality, and are intended to summarize the mortality experience of each occupational group. They express the number of deaths in the year for men of a particular occupation (aged 20–64 years) as a percentage of those *expected* in that year, had the sex and age mortality of a *standard period* been operating on the sex and age *population* of the occupation and year in question.

Standardized mortality ratios

Table 20.5 illustrates the method of calculation in a hypothetical example:

Table 20.5 Standardized mortality ratios in a population

1	*2*	*3*	*4*	*5*	*6*
				Products	
Age group	*Male population in the occupation (thousands)*	*Actual death-rate per 1000*	*Standard death-rate per 1000*	*Actual deaths (2) × (3)*	*Expected deaths (2) × (4)*
20–24	5	2.0	1.0	10.0	5.0
25–29	8	1.0	1.0	8.0	8.0
30–34	15	2.0	2.0	30.0	30.0
35–39	15	2.5	2.4	37.5	36.0
40–44	20	3.0	2.5	60.0	50.0
45–49	20	6.0	7.0	120.0	140.0
50–54	10	8.0	7.0	80.0	70.0
55–59	5	22.0	20.0	110.0	100.0
60–64	2	25.0	22.0	50.0	44.0
TOTALS	100			505.5	483.0

Note 1. The expected number of deaths, on the basis of the standard period is 483. Column 4 is the age-specific death-rate for all males in England and Wales, regardless of occupation.

Note 2. The actual number of deaths, during the year, is 505.5.

Note 3. The standardized mortality ratio would be:

$$\frac{505.5}{483} \times 100 = 104.7 \text{ per cent}$$

See the next section for further details.

Use and interpretation

Crude birth- and death- rates

These are useful, but only approximate measures which determine whether or not a population will increase or decrease in the long run. For a short-period comparison of fertility and mortality experience in *one particular area*, these rates may be useful, since changes in the age and sex structure of a region are usually slow to emerge.

Standardized rates

As previously explained, crude rates can be regarded as weighted arithmetic means, the weights being the actual numbers of people in each age group at the time.

Standardized rates use the same weights for each calculation, so that any differences in age structure between one region and another are eliminated. They are all assumed to have the same *standard* age structure, and for purposes of pure comparison it does not matter which particular standard, or set of weights, is employed. For comparisons over a long period, however, it is desirable that this standard population should be typical of the age and sex structure of the present-day population. If the standard population were abnormally old or young, as with Havarest or Youthville, the resultant rates could be misleading.

Life tables

Readers will notice from Table 20.2 the significant difference between males and females at all ages. The expectation of life from birth is over six years longer for females, and although this difference tends to diminish throughout life, it is always present, being 1.4 years for persons aged 80 years. The same influence can be seen in the numbers surviving to any specified age. Out of our original 10 000 births, only 2340 males remain at age 80 years, but there are 4495 females.

The implications of this factor can be important for manufacturers of consumer goods, who will find it more beneficial to cater for women than for men, particularly in the older age groups. Social services, old people's homes, etc., can also be geared to this lack of balance between the sexes.

310

Life tables are one of the devices used by the Registrar General to estimate the size of the future population, and particularly its age structure. The present tendency is for the proportion of older people of both sexes to increase, and this has important economic implications in the field of taxation, pensions, etc. Life assurance companies base their premiums on the data of these tables, and the calculation of annuity rates also depends upon them.

Reproduction rates

The gross rate is based upon impractical assumptions and is of little real value.

A net rate of *unity* implies that the current female population is replacing itself exactly. If it remains below unity, then at some future date a fall in the population will follow unless immigration or some similar factor makes up the deficit.

In effect, the net rate defines the trend of numbers per generation that would ultimately result from the indefinite continuation of the age-specific fertility and mortality rates from which it was calculated. It is only one way of allowing for the disturbing effects of an abnormal age distribution, and other abnormalities may also affect the number of births.

Recent experience has shown that the age-specific fertility rates of a particular year, or series of years, may be affected by factors such as the proportion who have *recently* married, and by fluctuations in the rate at which married couples build up their families. For example, a family may decide upon three children but may not space them out evenly throughout the reproductive years.

For many years ahead, the population will consist mainly of survivors of those already alive. As far as these people are concerned, the trend implied by current reproductivity is quite irrelevant.

Area comparability factors

The purpose of this measure is to adjust the various local crude death-rates to allow for differences in the age structure.

The death-rates obtained from columns 5 and 6 of Table 20.4 are sometimes called the *index* rates. These are 22.02 for Havarest, and 7.74 for Youthville. They measure the relatively favourable age distribution of an area, as regards mortality, compared with a national standard (12.6 in the table). The figure of 7.74 for Youthville, for example, means that its age distribution is more favourable than that of England and Wales (12.6). Other things being equal, it ought, therefore, to have a lower death-rate.

When this index is divided into the standard rate, the resulting *factor* will be greater than unity (1.63). When this, in turn, is multiplied by the crude rate for the same region (8.89) the effect will be to increase the crude rate. The reverse will apply for Havarest. In other words, the A.C.F. is to be compared with a figure of *unity* for the whole of England and Wales.

Areas with a factor greater than unity have more favourable age distributions, and the A.C.F. corrects for this—similarly, of course, for areas with an A.C.F. of less than one.

A useful measure or *ratio* is the local adjusted death-rate divided by the national (standard) rate. This gives a figure of:

$$\frac{11.54}{12.6} = 0.92 \text{ for Havarest}$$

$$\frac{14.5}{12.6} = 1.15 \text{ for Youthville}$$

These ratios can be compared directly, both with each other and with any other area in that particular year. The ratio for England and Wales is *unity*.

Hence, we conclude that Havarest is better than the national average, from the point of view of health, as well as being better than Youthville.

Standar-dized mortality ratios

These facilitate comparisons between the mortality of a particular occupation, and that of England and Wales as a whole.

From Table 20.5, this ratio is 104.7 per cent, indicating that this particular occupation is less healthy than the male population as a whole. A figure of less than 100 per cent indicates a *more* healthy occupation.

These ratios are calculated for 'all causes' of death, and also for specified causes, such as cancer, tuberculosis and various types of accident. The mortality rates for each disease are analysed by social class, by sex and by marital condition. The standardized mortality ratio is also used to compare regional differences as regards mortality.

Final note

The use and analysis of vital statistics is a major tool of government and business. Population structure and trends form the basis of planning in social services, education, housing, transport, and so on. Taxation policy and pensions are also governed by these factors.

In the business field, such information is of great importance. The large-scale chain stores and supermarkets must know whether a particular area will justify a new branch. This does not depend solely on population size, since the area might be changing. The age and sex distribution, birth- and death-rates, and the rate of change in such measures are all relevant. Youthville will support different types of shop from Havarest.

The manufacturer is equally concerned with trends in birth-rates, age and sex structure, etc. How will the influx of immigrants, with possibly different birth- and death-rates, affect national trends?

A high birth-rate is good for baby foods and clothes, prams, and toys. An ageing population needs more of the things used by older people. Pensioners have less to spend, and spend it differently.

Death rates on MINITAB

The examples calculated above (e.g. standardized death-rates and mortality ratios) can easily be replicated in MINITAB (and spreadsheets) by keying the basic data from the tables into the worksheet. We then perform column operations with the LET command.

Suppose, however, that we had not yet reduced the data to a tabular form like that of Table 20.1. We might, for instance, want to calculate the sex ratio in a firm from the individual level data in our personnel files. The variable 'SEX' in our MINITAB workfile of 250 rows (250 employees) is coded '1' for women, '2' for men. We want to calculate the sex ratio for the company:

$$\text{Sex ratio} = \frac{\text{total number of females} \times 100}{\text{total number of males}}$$

One way of doing this in MINITAB is to use UNSTACK:

```
MTB > UNSTACK 'SEX' into C100–C101;
SUBC > SUBSCRIPTS 'SEX'
```

This UNSTACK command copies the rows of the SEX variable to one of two new columns, C100 or C101. The SUBSCRIPTS subcommand indicates that the codes that direct where data are to be copied to are in the SEX variable itself. All rows of the SEX variable for the value '1' are copied to C100, all rows of value '2' to C101. Column C1 held 250 observations of 168 women and 82 men; column C100 now holds 168 rows of '1' and column C101 holds 82 rows of '2'. Check this with the INFO command if you are uncertain.

We complete the calculation with LET:

```
MTB > LET K1 = COUNT(C100)*100/COUNT(C101)
MTB > PRINT K1
```

We are given the value of the constant 'K1' as 204.878. This is our sex ratio.

21

Sources of statistics

The description of sources in this chapter is divided into two parts: published statistics, and private business statistics.

Published statistics

Statistics in this section, a selection from the vast range available, are described under six headings:

1. Population.
2. Labour.
3. Production.

4. Prices.
5. Trade
6. National Income.

Information and sources

Tables of suggested publications which deal with the statistics under each heading are given below:

Table 21.1 Population

Information	Publication (occurrence)	Source
Reports of the Census of Population (except 1941)	Census Reports (periodic)	Offices of Population Censuses and Surveys (OPCS)
Part 1 Deaths, death-rates, causes of death *Part 2* Number in populations, marriage and fertility rate. Number of electors (parliamentary and municipal) *Part 3* Tables, analysis and historical survey	Registrar General's Annual Statistical Review of England and Wales	OPCS
Number in population, births, marriages, deaths, including figures of sickness and industrial diseases	Population trends (quarterly)	OPCS
Births, deaths and infectious diseases in large centres	Monitor Series (monthly)	OPCS

Table 21.2 Labour

Information	Publication (occurrence)	Source
Total employees by sex and age, totals employed in various industries and by region	Annual Report of the Department of Employment	Department of Employment
Minimum or standard rates of wages and weekly hours of work. Summaries of collective agreements and statutory wage regulation orders	*Employment Gazette* (monthly) New Earnings Survey (annual)	Department of Employment
Industrial accidents, causes, nature, site of accidents, fatalities	Annual Report of Chief Inspector of Factories	Department of Employment
Monthly summary of employment, unemployment, unfilled vacancies, overtime and short-time working, rates of wages, stoppages of work. Special articles each month	*Employment Gazette* (monthly)	Department of Employment
Industrial diseases, poisoning, gassing, dust, fumes, radiation	Annual Report of Chief Inspector of Factories	Department of Employment
Trade union statistics	*Employment Gazette* (monthly)	Department of Employment

Table 21.3 Production

Information	Publication	Source
Index of Production	*Annual Abstract of Statistics* *Monthly Digest of Statistics*	Central Statistical Office (CSO)
Reports on the Census of Production	*Business Monitor* (quarterly)	Business Statistics Office
Iron and steel employment output at home and abroad. Import/export, fuel consumption, capacity, and types of products by the industry	Iron and Steel Industry (annual)	British Steel Corporation
Output stocks, employment, number of mines, types of product, investment (coal)	Report and Accounts (annual)	Department of Energy
Output, employment, capacity, investment, sales (electricity)	Report and Accounts (annual)	Department of Energy
Output, employment, capacity, investment, sales	Report and Accounts (annual)	Department of Energy

Note. Also censuses of production for Northern Ireland.
Also *Annual Report and Accounts for Northern Ireland, North of Scotland and South of Scotland*.
Also the *Digest of UK Energy Statistics*, which gives the main statistics on coal, electricity and gas (Department of Energy).

Table 21.4 Prices

Information	*Publication (occurrence)*	*Source*
Retail Price Index	*Employment Gazette* (monthly)	CSO
Producer, import and export prices	*Annual Abstract of Statistics Monthly Digest of Statistics*	CSO
Producer Price Index	Producer Price Index Press Notice (monthly)	CSO
Prices of agricultural supplies and products, month by month	*Agricultural Statistics UK* (Annual)	CSO
Indexes of share prices of groups of shares, e.g. capital and consumer goods, commodities, securities and stocks, and financial institutions	F. T.- Actuaries Share Indexes (daily)	*Financial Times* newspaper and the Institute of Actuaries in London and the Faculty of Actuaries in Edinburgh

Table 21.5 Trade

Information	*Publication (occurrence)*	*Source*
Reports on the Census of Distribution	Census Report (periodic)	Department of Trade and Industry
Monthly sales of retail establishments; turnover, staff	*Business Monitor* (quarterly)	Department of Trade and Industry
Quantities and values of commodities of export trade	*Overseas Trade Statistics of the UK* (Annual Supplement)	H.M. Customs and Excise
Quantities and values of commodities in monthly figures of export trade	*Monthly Review of External Trade Statistics*	Department of Trade and Industry
Import penetration and export sales ratios	*Business Monitor* (quarterly)	Department of Trade and Industry

Table 21.6 National Income

Information	Publication (occurrence)	Source
National Income and Expenditure (government and national accounts)	*Preliminary Estimates* (White Paper [annual])	CSO
Gross National Income, Expenditure and Product. Consumers' expenditure, personal incomes, company incomes, central and local government budgets, capital investment	*National Income and Expenditure* (Blue Book [annual])	CSO

The tables given above form a small part only of the mass of statistics published publicly and privately. Nevertheless, they afford a good selection of the main statistics that the student is likely to need. These statistics are repeated in simplified form in many other publications: e.g. nearly all these figures appear somewhere in *The Annual Abstract of Statistics*. The student should not neglect the quarterly issues of the *Business Monitor* series.

It is of vital importance that the student should make *personal contact* with several publications which contain these statistics. This is the only part of the course which this book cannot really provide. The following list is a requirement for the student's reading if any real acquaintance with the official sources of statistics is to be made.

The Annual Abstract of Statistics   **Publications**
(Compiled by the Central Statistical Office.)

Most of the statistics tabled above are to be found in this main reference book and, in most cases, comparative figures for earlier years are included. Its 350 tables generally give data for the last 11 years.

In particular, the notes and definitions contained in the book should be read and understood.

The Monthly Digest of Statistics
(Compiled by the Central Statistical Office.)

This is a monthly, abbreviated version of *The Annual Abstract of Statistics*, providing basic information on 20 subjects. The annual (January) *Monthly Digest of Statistics Supplement of Definitions and Explanatory Notes* gives definitions for items and units in the *Digest*.

Financial Statistics
(Compiled by the Central Statistical Office in collaboration with other government departments and the Bank of England. Published monthly.)

This is an important publication dealing with the key financial and monetary statistics of the UK.

The contents of this publication include financial statistics of the Exchequer and Central Government, local authority borrowings and public corporation accounts, banking statistics of the commercial joint-stock banks and the Bank of England. A wide range of financial statistics includes many dealing with discount houses, trustee savings banks, hire purchase finance companies, unit trusts, insurance, building societies, etc.

Also included are the income, finance and profits of companies, capital issues, and hire purchase credits granted. Many interest rates, security prices and yields, and local authority and building society mortgage rates are to be found in this publication.

A section dealing with overseas finance includes gold and foreign currency reserves, the balance of payments and foreign exchange rates.

A valuable end section deals with 'Notes and Definitions' at some length.

The annual *Financial Statistics Explanatory Handbook* contains comprehensive notes and definitions for the tables.

Economic Trends
(Compiled by the Central Statistical Office. Published monthly.)

Economic Trends is a compilation of all the main economic indicators liberally illustrated with charts and diagrams. The largest section gives the time series and graphs over the last five years or so. It is the primary publication for quarterly articles on national accounts and the balance of payments, as well as others commenting on and analysing economic statistics. The *Economic Trends Annual Supplement* contains longs runs of annual and quarterly figures for the key series of economic indicators.

Statistical News
(Compiled by the Central Statistical Office. Published quarterly.)

A comprehensive account of current developments in British official statistics and how they are compiled, this valuable publication contains many informative articles. A cumulative index is provided.

Regional Trends
(Compiled by the Central Statistical Office. Published annually.)

This publication deals with the breakdown of national statistics into their regional variations. It contains tables, sources and definitions on a broad range of economic, social and demographic topics.

Social Trends
(Compiled by the Central Statistical Office. Published annually.)

Social Trends provides analyses and breakdowns of statistical information on population, households and families, education and employment, income and wealth, resources and expenditure, health and social services, and many other aspects of British life and work.

The Guide to Official Statistics
(Compiled by the Central Statistical Office. Published occasionally.)

The first issue of this *Guide* was in 1976. This is an extremely important book and should be consulted by every student of statistics. It is really a bibliography of statistical sources, official and non-official, in the UK. All the important statistical series are covered and the *Guide* refers readers to the regular publications and also to special reports, articles, etc., with significant statistical content published over the last 10 years.

Each of its 16 subject areas is broken down into subsidiary areas. Thus, 'Labour', the fifth subject area, is divided into eight subsidiary area:

5.1 General sources of labour statistics.
5.2 Employment.
5.3 Qualified manpower.
5.4 Industrial training.
5.5 Unemployment, vacancies, placings and special employment schemes.
5.6 Industrial accidents and diseases.
5.7 Industrial relations.
5.8 Wage rates, earnings, labour costs, etc.

Each of these subsidiary areas is broken down into 'regular' and 'occasional' sources which lead the reader into more details of current figures and writings on the particular area.

Employment Gazette
(Compiled by the Department of Employment. Published monthly.)

The contents of each issue include:

1. A summary of the monthly statistics relating to labour: employment, unemployment, unfilled vacancies, overtime and short-time working, rates of wages, retail prices and stoppages of work.
2. Special articles related to the labour field.
3. Arbitration awards, notices, orders, statutory instruments, etc., published within the previous month.

The student should proceed carefully when reading this important publication, trying, for example, to define clearly and to personal satisfaction such topics as average wage rates, average weekly earnings, and hourly earnings and rates. Careful definition is also advisable when dealing with unemployment, because certain groups are not included in the count.

The Bank of England Quarterly Bulletin
This is the premier financial review and assessment of money and banking. Briefly, it may be broken down into:

319

1. General assessment (domestic and international) of recent economic and financial developments.
2. Valuable articles and speeches on money and banking.
3. A statistical appendix including government fiscal and monetary affairs.

No student of the economy can afford to neglect this publication.

United Kingdom National Accounts (Blue Book)
(Compiled by the Central Statistical Office. Published annually.)

This is the main source book on national income and expenditure. A brief analysis of the contents is given below:

1. Gross National Product.
2. Gross National Income.
3. Gross National Expenditure.

These three calculations are simply three different ways of looking at the money value of goods and services resulting from economic activity within the nation for one year.

For each calculation under these three headings the figures are analysed. For example, under *National Income* are to be found such entries as:

- Income from self-employment.
- Income from employment.
- Trading profits of companies.
- Surpluses and profits of public corporations and enterprises.
- Rents.
- Income from abroad.

Under *National Expenditure* are such entries as:
- Consumers' expenditure
- Public authorities' expenditure.
- Expenditure on capital goods (capital formulation).
- Imports—exports.

Under *National Product*, are such entries as:

- Agriculture, forestry and fishing.
- Insurance, banking and finance.
- Transport and communication.
- Public administration and defence.

These lists are by no means complete, and the student is advised to consult the actual tables of recent date to examine the complete lists, to note the figures for recent dates of the main items, and to become familiar with the main definitions.

Many other tables are given in the book, and these are largely derived from the main tables above. For the most part, they are more highly detailed analyses of the sections and items already mentioned. For example:

- Breakdown of the capital and current expenditures of central and local authorities.
- Distribution in class intervals of personal incomes before and after tax.
- Gross capital formation at home, analysis by sector, industry and type of asset; also on investment in stocks, and 'work in progress' by sector and industry.

The *Blue Book* contains comparative figures in most of its tables for the previous years.

United Kingdom Balance of Payments (Pink Book)
(Compiled by the Central Statistical Office. Published annually.)

This is the basic reference book for balance of payments statistics and the counterpart and complement to the Blue Book. It presents information for the last 11 years or more on the visible and invisible trade of the UK with the rest of the world.

Business Monitor
(Published quarterly for the Business Statistics Office by HMSO.)

This important publication deals with statistics especially of interest to the businessman or woman such as production, service and distributive series.

Euro-statistics
(Published occasionally by the Statistical Office of the European Community, Luxembourg.)

The importance of the UK's membership of the European Community makes this compilation of statistics essential reading for the student and the exporter. The 10- part statistical reviews cover government, labour, production, prices, finance and trade generally.

European Marketing Data and Statistics
This detailed manual is published by Euromonitor plc of Turnmill Street, London, and the 27th Edition was published in 1992. It contains a wealth of statistics relating to the European Community on population, finance and banking, import—exports, labour, production, energy, defence, pollution, consumer expenditure, retail data, advertising, housing, education, agriculture, transport, tourism and travel.

321

Consumer Europe

The first Edition was published in 1976, the 8th Edition in 1991. It uses extracts from the previous publication but includes forecasts to 1994.

Note on the standard industrial classifications

It was emphasized in Chapter 1 that statistics are comparable only when they are collected on a similar basis. This implies that the terms and definitions used are standardized.

Published statistics are collected from a wide variety of sources, and are issued by many different government departments. The Standard Industrial Classification (SIC) was drawn up to ensure that the various government statistics should be capable of comparison.

The classification is based on industries, not occupations, and the unit is the *establishment*, i.e. the whole of the premises. Thus, a tobacco factory might include such activities as selling, packing, transport and a canteen, as well as manufacturing. All these would be classed under the one heading.

The first Standard Industrial Classification for the UK was issued in 1948. It was revised in 1958 and 1968. The latest system of classification was issued in 1980 and includes all establishments according to industry covering all economic activity.

There are 10 broad divisions as follows:

0. Agriculture, forestry and fishing.
1. Energy and water supply industries.
2. Extraction of minerals and ores other than fuels; manufacture of metals, mineral products and chemicals.
3. Metal goods, engineering and vehicles industries.
4. Other manufacturing industries.
5. Construction.
6. Distribution, hotels and catering; repairs.
7. Transport and communication.
8. Banking, finance, insurance, business services and leasing.
9. Other services.

These 10 divisions are then divided into 60 classes. These are divided into 222 groups, which are subdivided into 334 activity headings. For example, Division 4 'Other Manufacturing Industries' is divided into 8 classes, 50 groups, and 91 activity headings.

Economic Briefing

These publications by the Information Division of the Treasury replaced in 1991 that department's previous *Economic Progress Reports*.

They give similar reports and statistics supported by graphs, bar charts, pie charts, and tables, and they contain special features (recently on the Stock Exchange, the Budget, the World Economy) as well as a diary of economic events.

The Briefings are free on application to: *Economic Briefing* (Distribution), Central Office of Information, HM Treasury, Parliament Street, London, SW1P 3AG. They may be freely copied for educational use.

National Westminster Bank Quarterly Review

This is a quarterly magazine, published by the National Westminster Bank, which contains several articles, one usually dealing with the agricultural scene, and current notes of topical interest.

Barclays Review

This is also a quarterly magazine, published by Barclays Bank, containing several articles usually biased towards the overseas economic scene. It is an excellent publication for the statistician, because it has a 'Statistical Section' containing many tables on numerous economic topics divided into 'Manpower, Production, Trade, Prices and Wages, Banking and Finance, Overseas Statistics'. Accompanying the review is usually to be found an extremely handsome coloured insert with several unusual kinds of diagram and symbol illustrations.

This short list of free publications certainly does not complete the range available to the student. Many of the larger firms issue house magazines and other helpful literature on request. Students of economic and business statistics cannot fairly complain that the necessary material is beyond their reach either in terms of cost or availability.

Finally, here are three other publications of which the student should take note, besides the 'city columns' of at least one better class newspaper:

The Financial Times (daily)

This is the specialist for economists and businessmen and women. It contains much statistical information in varied forms—charts, diagrams, graphs and tables. It is the leading newspaper for day-to-day economic and business articles and special occasional articles. It is also the leading source for stock and share prices and statistics, and it contains the valuable 'F.T.-Actuaries' share indexes.

The Economist (weekly)

This is the specialist weekly for the economist and those in business. It contains economic and business articles which give information about a large number of countries. Many statistical items are obviously found throughout its pages, but it also has a special statistical section on 'Money and Exchanges' and 'Stock Prices and Yields'.

The Banker (monthly)

This magazine is not restricted to the banker in its articles, nor is it restricted to this country. A special section entitled 'Financial Statistics' appears in each issue, and this consists of tables on banking, government and external monetary statistics. In most tables, comparative figures for earlier years are given.

323

For each heading and subheading, there is an exact and detailed definition of the activities that are included. These are so worded that all forms of industrial activity must fall within one or other of these categories.

The value of the SIC to the compilers of official statistics is obvious. It ensures the standardization of terms and definitions, so that all government departments are working on a comparable basis. The SIC numbers are quoted, when relevant, in Published Statistics, as a guide to their use and interpretation.

Private business statistics

The statistician in business

Figures, tables and graphs cost money to produce, and the statistician will be tested, like any other factor in production, according to the usefulness of the results produced. The function of a statistician is to aid efficient management, and the task is to present management with clear and concise summaries of the firm's business activities which will serve as a basis for policy, decision and action. The main work will be to supply figures, tables, diagrams, charts and graphs in the form of periodic returns or as business reports. On occasions the statistician may be called upon to design, to management requirements, forms and questionnaires which will bring in the kind of figures that can be used.

Different sizes of firm will need their own special degrees of statistical work. Smaller firms will often manage with very few statistics. One can hardly imagine a sweets and tobacco retailer with much statistical work to do, whereas the Department of Employment, without several specialist statistical departments, is unthinkable.

Different kinds of firm will be interested in different kinds of figures. The statistics in a travel agency will bear more heavily on the advertising and marketing side than those of a firm manufacturing steel girders. Statistics of stock turnover will be more apparent in egg production than in ship building. Average costs of labour and labour turnover statistics will figure more prominently on the books of a market gardener than in the records of an atomic power station.

Useful statistical data

A business unit of medium size will be interested in three main kinds of statistics:

1. *Published—general* These statistics will relate to the movements and change in national economic series. All businessmen and women will be interested in changes in interest rates, hire-purchase debt, stock and share prices, regional unemployment, etc.
2. *Published—particular* These are the statistics of a particular industry. If, for example, the business is shipbuilding, the interest in particular statistics might include:

   (a) *Freight rates*—a guide to the general profitability of passenger and freight ships.

(b) *Merchant shipping output*—a guide to the prosperity of the industry.

(c) *Export trade (by country and commodity)*—a guide to the kind of cargo ship to build and the most profitable runs.

(d) *Passenger movements*—see (a).

All these statistics, together with many more, could be gathered from the publications mentioned in this chapter.

3. *Private statistics*   These are statistics within the firm which can be compiled under the control of the businessman or woman.

It is possible to describe the mechanism of any firm under three general headings. The table below shows these three headings with the kind of statistics which might be found under each heading:

Table 21.7 Kinds of statistical data

Production	Marketing	Administration
Quality control	Market research	Internal audit
Labour turnover	Sales graphs	Borrowing and loans
Progress charts	Advertising	Bad debts
Stocks and inventories	Orders received and	Management ratios
Wage rates	fulfilled	Overhead expenses
Accident rates	Selling prices	Personnel and welfare
Sickness rates	Expenses of salesmen	Training system
Absenteeism rates	Distribution,	Government requirements
Unit costs	methods and costs	(payroll tax, industrial
Output charts	Export licences	training board, income
Purchasing costs of raw		tax, census reports)
materials		
Stock turnover		

Note.   Some of these statistics may need further explanation. For example, most rates can be derived as percentages of a ratio; thus, 'accident rate' could be the result of the formula:

$$\frac{Number\ of\ accidents\ in\ a\ period}{Average\ number\ of\ workers\ in\ period} \times 100$$

The management ratios referred to above are also simple rates. As in the case of accident rates, they are simple figures for quick reference but are none the less important to management's overall picture. Some of the more important ratios might be:

$$\frac{Current\ assets\ to\ date}{Current\ liabilities\ to\ date} \times 100$$

$$\frac{Sales\ during\ period\ (value)}{Average\ fixed\ assets\ in\ period,\ etc.} \times 100$$

Many other management ratios may be calculated, but their formulation is the job of the cost accountant rather than the statistician.

The little 'Overhead expenses' includes most expenses (other than fixed plant) which do not vary with output, e.g. fuel, light, rates, water, stationery, secretarial work.

Kinds of statistical methods

AVERAGES

Averages are the most widely used methods in a business unit. Examples are many, but the arithmetic mean will be used for most data; e.g. in rates of wages, overtime, earnings, accidents, sickness, absenteeism, costs per unit of product, and labour turnover for male, female and trainee labour in the separate grades of skill. The mode is also useful where modal sizes and types are common in stocks and inventories, stock turnover, and sales of standard sizes and models.

GRAPHS

Graphs are particularly useful in marketing and production. An obvious example is sales. Many graphs will be linked with times series analysis in order to determine seasonal trends and cycles. Other examples are output, purchases and costs of raw materials, order books, progress charts, unit costs and selling prices, stock graphs, bad debts, and borrowing and loans. Certain special graphs will be used in quality control, and semi-logarithmic graphs may be used to compare series of different kinds, e.g. orders received and advertising budget.

TABULATION

This technique is, of course, the basis of graphs, and will be used where the latter are required. Tables may be used as an alternative to graphs, and they usually find a prominent place in the business reports and returns which the statistician submits to management.

TIME SERIES

As noted above, this is an important part of the information a businessman or woman requires in order to compare present performance with the past, and also to check on seasonal movements—e.g. when to buy, when prices are lowest, when to sell, when prices are highest, and how to invest in the expectation of future performance by the extrapolation method.

QUESTIONNAIRES AND PERSONAL INTERVIEWS

In a larger firm these methods may find a place when the statistician requires information from the public about its reaction to the product, and from the worker regarding his reaction to the firm. Thus, the possible fields might be market research, and personnel and welfare.

DIAGRAMS AND SYMBOLS

Once again the statistician might use these methods in respect of the public and the worker. Help with advertising presentation may be required by the advertising copywriters' department. Production posters, target incentive displays and information sheets may be required on the factory floor.

SAMPLING

The student can grasp some idea of the method from Chapters 2, 14 and 16. This should be sufficient to realize that the sampling method will be used frequently in checking work where total checks would take too long or be wasteful. The sample check of the internal audit is a good example of such use. Sampling is the basis of quality control where continuous checks must be made on a standard machine product to ensure that batches are made to specification sizes within certain tolerance margins. If the firm is large enough to conduct market research surveys, it is possible that sampling techniques will be used here as well.

Conclusions

As was noted previously, there is a great variation in the possible work of a statistician in business. This is partly due to the differing sizes of firms, partly to the different requirements of each kind of firm, and partly to the fact that no standard method of statistical working exists in every firm. The student will no doubt be aware of the similar discrepancy between bookkeeping theory and actual practice. Even less uniformity among firms exists in statistics than in bookkeeping. Because of this, we have not entered into the details of statistical practice in business. Each firm evolves its own particular methods, and the wide variations would form no general rules for students at this stage of their work.

Typical examination questions

Chapter 2

1. A detergent manufacturer wishes to obtain information on a sample basis about the use of the company's product in a large town. Draft a short questionnaire with instructions for its completion to be sent by post to a group of families in the town. How might the families be selected for inclusion in such a survey? As this is an important enquiry the expenditure involved is not a major consideration.

2. Of what value are the results of the Censuses of Population
   (a) to a businessman.
   (b) to a government body concerned with long-term planning?
   Illustrate your answer with reference to the last Census of Population of Great Britain.

Chapter 5

1. Construct a blank table to show absenteeism among the employees of an industrial establishment according to the length of absence ('under three days', 'three days and under one week', 'one week and under two weeks', 'two weeks and over'), classifying the absentees by sex, by civil condition ('single' or 'married') and by age ('under 18 years', '18 years and under 50 years', '50 years and over').
   Note. The table should be designed to apply to *one* month only. Totals and subtotals may be included.

2. You are asked to make a report on the length of service and efficiency of a set of computers. Draw up a tabulated statement giving the following information:
   (a) *Age of machine*—under one year, one year, and under two years, etc.
   (b) *Type of machine*—Amstrad, IBM, etc.
   (c) *Condition of machine*—serviceable, slight repairs, etc.

Chapter 6

1. The following figures relate to the sales and profits of a well-known multiple store:

Year of review	Sales		Profits	
	£ million	Per cent increase on previous year	£ million	Per cent increase on previous year
First	107	14.0	11.2	13.0
Second	118	10.5	12.4	10.7
Third	125	6.0	14.7	18.6
Fourth	130	4.0	16.2	10.2
Fifth	134	3.0	17.1	5.5

Draw a suitable graph or diagram to bring out the relationship between sales and profits.

2. (a) State the advantages and the disadvantages of diagrammatic presentation of statistical information.
   (b) Describe the type of graph or diagram suitable for presenting the following information:
      (i) The sales in *four* areas, of *three* products, for the last *two* years.
      (ii) The total expenditure of a bus company during a given year, showing the proportions due to wages, fuel, maintenance, tyres and depreciation.
      (iii) Broadcast receiving licences on 30 June, over the last 10 years and divided into sound and television.

Chapter 7

1. The final results in mathematics of 80 students are recorded in the following table:

68	84	75	82	68	90	62	88	76	93
73	79	88	73	60	93	71	59	85	75
61	65	75	87	74	62	95	78	63	72
66	78	82	75	94	77	69	74	68	60
96	78	89	61	75	95	60	79	83	71
79	62	67	97	78	86	76	65	71	75
65	80	73	57	88	78	62	76	53	74
86	67	73	81	72	63	76	75	85	77

Construct a frequency distribution from the above data (showing your working); comment on your choice of class intervals and plot a histogram and a frequency polygon.

2. The following table shows the distribution of marks gained by students in the same statistics examination. The papers, however, were marked by different examiners. If they pass the same proportion of candidates, estimate by means of cumulative percentage frequency curves (on the same axes) the pass mark for examiner A if the pass mark for examiner B is 55. Assume that the average ability and the scatter of ability of the two groups are approximately the same and that each student has done better work than those given fewer marks.

Marks	Examiner A	Examiner B
0–10	1	0
11–20	4	2
21–30	12	16
31–40	47	55
41–50	64	62
51–60	42	66
61–70	18	59
71–80	8	28
81–90	2	8
91–100	2	4

3. On ordinary graph paper, construct a semi-logarithmic chart of the UK exports of electrical machinery in the three categories given. Explain how the chart you have constructed differs from an ordinary one. Show clearly the figures you obtain to plot on your graph.

	Electrical industry (UK Exports of electrical machinery—£ million)		
Year	Total	Commonwealth	European Community
Year 1	276.4	163.2	26.5
Year 2	285.2	164.9	28.0
Year 3	293.5	161.7	34.9
Year 4	318.5	156.7	47.1

Chapter 8

1. (a) Explain by means of your own examples:
   (i)   Absolute errors.
   (ii) Relative errors.
   (iii) Unbiased errors.
   (iv) Biased errors.
   (b) A manufacturer is planning to market a new article. It is estimated that, if the article is priced at 50p, 10 000 (to the nearest 1000) would be sold in the first year. Expenses are estimated to be as follows:

   Wages   £1000 (to the nearest £100)
   Materials   £2000 ($\pm$10 per cent)
   Overheads   £850 (to the nearest £50)

Calculate:
 (i) The total estimated profit.
(ii) The estimated profit per article.
In each case, give the limits of the error of the estimate.

2. The following figures are approximations only, their limits of error being as shown. Carry out the required operations and indicate in each of your answers the limits of error to which it is subject:
(a) Add 340 ($\pm$5) to 500 ($\pm$25).
(b) Subtract 18 ($\pm$2) from 90 ($\pm$10).
(c) Multiply 500 ($\pm$50) by 10 ($\pm$1).
(d) Divide 125 ($\pm$5) by 5 ($\pm$1).

Chapter 9

1. (a) State between what limits the following product lies if the numbers are rounded off to the given significant figures:

$$9.73 \times 2.41$$

   (b) Calculate:
      (i) the arithmetic mean
      (ii) the median
      (iii) the mode
      of the following numbers:

      10, 14, 22, 16, 15, 14, 15, 13, 14, 17

2. Obtain by calculation the arithmetic mean and median of the following distribution. Discuss the relative advantages and disadvantages of the two measures in the example.

Age of Birmingham taxi drivers	
Age in years	Per cent of all drivers (rounded to nearest whole percentage)
Under 35	13
35 and under 40	11
40 and under 45	15
45 and under 50	16
50 and under 55	14
55 and under 60	10
60 and under 65	9
65 and under 70	7
70 and under 75	3
75 and over	2
TOTAL	100

Would you have been surprised if the percentage columns had not totalled 100 per cent? Why?

3.

England and Wales	
Age of mother (years)	Number of live births per 1000 women
15–19	31
20–24	158
25–29	161
30–34	94
35–39	46
40–44	12
45–49	1
TOTAL	503

(a) Calculate, using a short a method, the average age of mothers giving birth to a child.
(b) Represent the data graphically by means of a histogram and insert the mean age in your diagram.

Chapter 10

1. Two members of a sales staff have the following bonus earnings in pounds sterling per month:

Member	Jan.	Feb.	Mar.	Apr.	May	June
A	12.3	12.27	12.65	12.88	12.06	12.5
B	12.3	12.19	12.38	12.46	12.08	12.3

Member	July	Aug.	Sept.	Oct.	Nov.	Dec.
A	12.69	12.92	12.28	12.55	12.82	12.08
B	12.2	12.55	12.18	12.37	12.49	12.1

Calculate the arithmetic mean and the standard deviation of each set of results and state, giving your reasons, which member, A or B, you would prefer to work for you.

2. (a) What basic characteristic of a frequency distribution are represented by:
   (i) The median?
   (ii) The mean deviation?
   (b) Calculate the mean deviation of the following distribution of weights:

Weight in lb.	Number of Boxes
10 and under 12	2
12 and under 14	9
14 and under 16	20
16 and under 18	25
18 and under 20	24
20 and under 22	15
22 and under 24	5
TOTAL	100

3. (a) Explain and illustrate graphically the relationship between the mean, median, and mode in
   (i) A positively skewed frequency distribution
   (ii) A negatively skewed distribution.
   (b) Calculate the coefficient of skewness for a frequency distribution with the following values: mean = 10; median = 11; standard deviation = 5. What does your answer tell you about this frequency distribution?

Chapter 11

1. The following data show the present maintenance costs and age of a sample of eight similar machines used in a factory:

Machine	A	B	C	D	E	F	G	H
Age of machine (years)	1	3	4	4	6	7	7	8
Maintenance cost (£)	£200	£550	£650	£800	£1150	£1100	£1300	£1500

(a) Calculate and plot the regression line of maintenance cost on machine age.

(b) Use the regression line to predict the expected maintenance cost for a machine aged 10 years. Comment on the accuracy of your prediction.

2. (a) What is meant by a coefficient of correlation? How does correlation analysis differ from regression analysis?

(b) The following table gives X, the number of radio sets (in thousands), and Y, the number of television sets (in thousands), sold in Camford over the last nine months:

X	140	150	130	100	90	80	90	120	130
Y	270	350	340	240	190	170	150	180	170

Plot these data and discuss the extent of correlation between X and Y

3. The following data refer to travelling expenses and duration of trips for consecutive trips made by sales engineers of a large firm:

Trip number	1	2	3	4	5	6	7	8
Expenses (£)	43	24	30	26	17	29	28	29
Duration of trip (days)	9.0	3.5	6.5	4.0	2.0	5.0	3.0	3.5

(a) Plot these data on a suitable graph.

(b) Estimate the regression equation for predicting total expense from duration of trip.

(c) Use the regression equation to estimate the average expense for a trip of duration five days.

Chapter 12

1.

	Company sales (£ thousands)			
Year	1st quarter	2nd quarter	3rd quarter	4th quarter
1			115	105
2	141	156	129	130
3	148	184	140	176
4	224	248	196	139
5	207	240		

(a) Smooth the above time series by means of a moving average.
(b) Plot both the original series and the smoothed series on the same graph.
(c) Estimate the trend of figures for sales in the third and fourth quarters of Year 5, and comment on the reliance which can be placed upon these estimates.

2. State what you understand by the seasonal cycle in a time series. Estimate the seasonal element in the data below and illustrate the cycle with a diagram.

	Year 1	(£ millions) Year 2	Year 3
1st quarter	3118	3191	3376
2nd quarter	3288	3468	3603
3rd quarter	3368	3494	3614
4th quarter	3618	3767	3822

Chapter 13

1. A firm wishes to follow the relative movement of the prices of various raw materials which it uses. It decides to construct a simple index number starting from January Year 3 and based on average prices from Year 1.
   (a) From the following data, calculate an index for the month of January Year 3.

Raw materials	Price (£ per ton) Average, Year 3	January Year 1	Weight
A	16	19	5
B	24	25	1
C	13	18	3
D	8	9	6
E	12	14	4
F	4	8	3

   (b) What are the main points the firm should bear in mind when planning the construction of this type of index number?

2.

	November Year 1		March Year 4	
	Average price (new pence) per lb	*Average quantity purchased per week*	*Average price (new pence) per lb*	*Average quantity purchased per week*
Rump steak	$70\frac{1}{2}$	1 lb	98	$\frac{3}{4}$ lb
Potatoes	2	2 lb	4	$2\frac{1}{2}$ lb
Tomatoes	$16\frac{1}{2}$	$\frac{1}{2}$ lb	20	$\frac{1}{2}$ lb
Cauliflower	10	$1\frac{1}{2}$lb	15	1 lb

From the above data find the following price indexes for March Year 4, with November Year 1 = 100:

(a) Unweighted index.

(b) Laspeyre index (base year weighting).

(c) Paasche index (current year weighting).

Comment on your results.

Chapter 14

1. A survey obtained the following public opinion replies to the question, 'If there were a general election tomorrow, how would you vote?'

For the Government	480
Against the Government	560
TOTAL REPLIES	1040

Three months later the replies were:

For the Government	440
Against the Government	640
TOTAL REPLIES	1080

Do the results suggest any significant change in political opinion over the period? Would you be surprised, if six months after the second survey, the Government won the General Election?

2. Among patients having a certain disease, in Hospital A 20 per cent died and in Hospital B 8 per cent died. The number of patients in the two hospitals were:

	A	B
Men	21	57
Women	84	18

Would you conclude that the patients received better care in Hospital B than in Hospital A? What additional figures would Hospital A use to defend its record?

3. A given type of aircraft develops minor trouble in four per cent of flights. Another type of aircraft on the same journeys develops trouble in 19 out of 150 flights. Investigate the performance of the two types of machines and comment on any significant difference.

Chapter 15

1. Three letters are typed and addressed to three different people, together with the corresponding envelopes. If the letters are put in the envelopes at random, what are the chances that 0, 1, 2 or 3 letters will be in the right envelopes?

2. Define probability.
   The probability that a man will be alive in 25 years time is 3/5 and the probability that his wife will be alive in 25 years time is 2/3.
   Find the probability that:
   (a) Both will be alive.
   (b) Only the man will be alive.
   (c) Only the wife will be alive.
   (d) At least one will be alive.

3. Define probability. What is mathematical expectation?
   (a) If a woman purchases a raffle ticket, she can win a first prize of £500 or a second prize of £200 with probabilities of 0.001 and 0.003. What should be a fair price to pay for the ticket?
   (b) In a business venture, a woman makes a profit of £300 with probability of 0.6 or takes a loss of £100 with probability of 0.4. Determine her expectation.

4. A process produces batches of items in which the number of defectives is known to be normally distributed with mean 10 and standard deviation 2.
   (a) Sketch the shape of this frequency distribution, indicating its basic characteristics.
   (b) State the probability that the defectives in a given batch will be:
      (i) less than 6.
      (ii) more than 12.
      (iii) between 8 and 14.

Chapter 16

1. The following data relate to measurements of components produced by two different machines. Is there any significant difference between the output of the two machines?

Machine	Number measured	Average size	Standard deviation
A	1200	31 in.	5 in.
B	1000	30 in.	4 in.

2. A manufacturer aims to make electric light bulbs with a mean working life of 1000 hours. He draws a sample of 20 from a batch and tests it. The mean life of the sample bulbs is 990 with a standard deviation of 22 hours. Is the batch up to standard?

3. In order to find whether intensive national advertising of milk affected sales in a particular area, two systematic samples of families were taken from milk rounds' books, one before and one after the campaign. With the help of means and standard deviations, determine whether the campaign was successful, ignoring all other factors influencing the pattern of milk sales.

Number of pints bought per week	Number of families in:	
	Sample 1 Before the campaign	Sample 2 After the campaign
Up to 5	77	81
$5\frac{1}{2}$–$10\frac{1}{2}$	151	128
11–16	213	194
$16\frac{1}{2}$–$21\frac{1}{2}$	254	295
22–27	428	372
$27\frac{1}{2}$–$32\frac{1}{2}$	230	323
33–38	92	146
$38\frac{1}{2}$ and over	45	79
TOTALS	1490	1618

Chapter 17

1. The following measurements were taken in the course of routine quality control checks at a factory. Six items were measured every hour:

Time	Results
10 a.m.	95, 96, 100, 103, 98, 92
11 a.m.	102, 99, 91, 100, 95, 89
12 noon	96, 101, 99, 94, 91, 103
1 p.m.	100, 90, 98, 102, 101, 95
2 p.m.	97, 88, 100, 103, 99, 101
3 p.m.	103, 101, 105, 103, 93, 98

(a) Draw 'mean and range' quality control charts with the following limits:

Mean chart—warning limits: 98.5 and 93.5
—action limits: 100 and 92

Range chart—warning limit: 14
—action limit: 15.5

(b) Enter the information for the day's production on your charts.
(c) How are the values for the limits obtained in practice?

2. Using the chi-squared (χ^2) distribution as a test of significance, test the statement that the number of defective items produced by two machines, as shown in the following table, is independent of the machine on which they were made.

Machine	Machine output		Total
	Defective articles	Effective articles	
A	25	375	400
B	42	558	600
TOTALS	67	933	1000

Use the 0.01 level of significance.

3. Two factories, using materials purchased from the same supplier and closely controlled to an agreed specification, produce output for a given period classified into three quality grades as follows:

Factory	Quality grade Output in tons			Total
	A	B	C	
X	42	13	33	88
Y	20	8	25	53
TOTALS	62	21	58	141

(a) Do these output figures show a significant difference at the five per cent level?

(b) What hypothesis have you tested?

Chapter 18

1. The following figures come from the Report on the Census of Production:

Textile machinery and accessories	
Establishments Numbers	Net output (£ thousands)
48	1406
42	2263
38	3699
21	2836
26	3152
16	5032
23	20385
TOTALS 214	38773

Analyse this table by means of a Lorenz curve and explain what this curve shows. What other information is available from the Census of Production Report?

2.

	Jan.	Feb.	Mar.	April	May	June	July	Aug.	Sept.	Oct.	Nov.	Dec.
Monthly production of the 'XL' Company, Year 1/Year 2 (thousand units)												
Year 1	400	410	360	400	420	450	430	380	410	450	460	430
Year 2	420	430	410	380	410	470	450	400	460	490	510	480

Calculate and plot a Z chart for Year 7.

Chapter 19

1. Quotations have been received from three firms for the purchase of a new machine cost price £16 000, the only difference being the terms of settlement which are as follows:

(a) Firm X—£8000 on delivery and £2000 payable at the end of each of the following four years.

(b) Firm Y—£4000 at the end of each year for the next four years.

(c) Firm Z—£5000 payable on delivery, £3000 at the end of the second year and third year and £5000 at the end of the fourth year.

The average cost of capital is 10 per cent.

(i) Show by means of a statement using the net present value (NPV) method which quotation should be accepted.

Present Value of £1 (10 per cent)	
Year 1	0.909
Year 2	0.826
Year 3	0.751
Year 4	0.683

(ii) State *two* advantages of using discounted cash flow methods when assessing capital projects.

(iii) State *two* difficulties met with in practice when applying such techniques.

2. An expanding company regularly examining opportunities for new investment, is considering the following two alternative projects. It is found that when the two sets of cash flow are discounted at 15 per cent per annum, they both result in a nil present value. The cash flows, after multiplying by the discount factors, are as follows:

Year	Project A Cash in	Project A Cash out	Project B Cash in	Project B Cash out
	£	£	£	£
1		30 000		26 000
2	9000		1000	
3	7500		3500	
4	6200		5000	
5	7300		4250	
6			3600	
7			3050	
8			5600	
TOTALS	30 000	30 000	26 000	26 000

You are required to state the principles on which a choice should be made between the two projects, making assumptions where necessary to illustrate your answer.

3. You are given the following information:

Investment at end of year 0: £31.7
Life of project: four years
Expected cash flow: £10 a year for four years

Find, by use of discounted cash flow method, the rate of interest that will repay the cost of the capital investment. What amount of interest will have been earned?

Chapter 20

1. Write brief notes on *four* of the following terms:
   (a) Vital statistics.
   (b) Population pyramid.
   (c) Crude birth rate.
   (d) Standardized death rate.
   (e) General Register Office.

2. From the vital statistics of a North Midlands town, the following data were extracted:

| | Age group (years) | | | |
	0 and under 10	10 and under 40	40 and under 60	60 and over
Population (thousands)	16	48	12	4
Deaths	212	247	291	337
Standard age distribution	150	400	300	150

   (a) Calculate the crude and standardized death-rates.
   (b) Explain with reasons why the crude death-rates are inadequate.

3. (a) What is standardized death-rate?
   (b) Given the following data, prepare a report estimating the number of extra pupil places which were required in the first year classes of state primary schools in England in each of the following years:

Live births in England and Wales (thousands)		
Year	Total	Wales alone
1	739	42
2	750	42
3	782	44
4	803	44
5	840	45
6	856	47

What further data would you like to have, and how would you use such data?

Chapter 21

1. Describe the nature of the statistical information concerning the UK which is published in official sources on any *four* of the following:
   (a) Distribution of manpower.
   (b) Production of iron and steel.
   (c) Building and construction.
   (d) Retail distribution establishments.
   (d) Retail prices.
   (f) Wages and salaries.

2. Write notes on *three* of the following:
   (a) The Census of Population.
   (b) Standard Industrial Classification.
   (c) Retail sales indexes.
   (d) The Census of Distribution.
   (e) Imports and exports price indexes.

3. What information does the Government publish in relation to labour statistics?
   Which are the main publications which contain these figures?

Appendixes

Appendix 1

**Table of
the normal
curve**

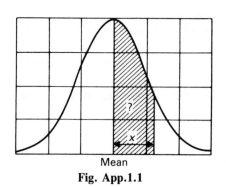

Mean

Fig. App.1.1

X	Area	X	Area	X	Area
0.0	0.0000	1.6	0.4452	3.0	0.4987
0.1	0.0398	1.7	0.4554	3.09	0.4990
0.2	0.0793	1.8	0.4641	3.1	0.4990
0.3	0.1179	1.9	0.4713	3.2	0.4993
0.4	0.1554	1.96	0.4750	3.3	0.4995
0.5	0.1915	2.0	0.4772	3.4	0.4997
0.6	0.2257	2.1	0.4821	3.5	0.4998
0.7	0.2580	2.2	0.4861	3.6	0.49984
0.8	0.2881	2.3	0.4893	3.7	0.49989
0.9	0.3159	2.4	0.4918	3.8	0.49993
1.0	0.3413	2.5	0.4938	3.9	0.49995
1.1	0.3643	2.58	0.4951	4.0	0.49997
1.2	0.3849	2.6	0.4953	4.1	0.4998
1.3	0.4032	2.7	0.4965	4.2	0.49999
1.4	0.4192	2.8	0.4974		
1.5	0.4332	2.9	0.4981		

Areas of normal curve

X is the distance from the mean, measured in standard deviations i.e. $\dfrac{value\text{-}mean}{S.D.}$

Appendix 2

Table of χ^2

v	0.9	0.5	0.1	0.05	0.01	0.001	v	0.9	0.5	0.1	0.05	0.01	0.001
1	0.016	0.45	2.71	3.84	6.63	10.83	16	9.31	15.34	25.54	26.30	32.00	39.25
2	0.21	1.39	4.61	5.99	9.21	13.82	17	10.09	16.34	24.77	27.59	33.41	40.79
3	0.58	2.37	6.25	7.81	11.34	16.27	18	10.86	17.34	25.99	28.87	34.81	42.31
4	1.06	3.36	7.78	9.49	13.28	18.47	19	11.65	18.34	27.20	30.14	36.19	43.82
5	1.61	4.35	9.24	11.07	15.09	20.52	20	12.44	19.34	28.41	31.41	37.57	45.32
6	2.20	5.35	10.64	12.59	16.81	22.46	21	13.24	20.34	29.62	32.67	38.93	46.80
7	2.83	6.35	12.02	14.07	18.48	24.32	22	14.04	21.34	30.81	33.92	40.29	48.27
8	3.49	7.34	13.36	15.51	20.09	26.13	23	14.85	22.34	32.01	35.17	41.64	49.73
9	4.17	8.34	14.68	16.92	21.67	27.88	24	15.66	23.34	33.20	36.42	42.98	51.18
10	4.87	9.34	15.99	18.31	23.21	29.59	25	16.47	24.34	34.38	37.65	44.31	52.62
11	5.58	10.34	17.28	19.68	24.73	31.26	26	17.29	25.34	35.56	38.89	45.64	54.05
12	6.30	11.34	18.55	21.03	26.22	32.91	27	18.11	26.34	36.74	40.11	46.96	55.48
13	7.04	12.34	19.81	22.36	27.69	34.53	28	18.94	27.34	37.92	41.34	48.28	56.89
14	7.79	13.34	21.06	23.68	29.14	36.12	29	19.77	28.34	39.09	42.56	49.59	58.30
15	8.55	14.34	22.31	25.00	30.58	38.70	30	20.26	29.34	40.26	43.77	50.89	59.70

v = The number of degrees of freedom
P = The probability of exceeding the tabular value of χ^2 in random sampling

(Abridged from Table 8 of *Biometrika Tables for Statisticians, Vol 1*, by kind permission of the Trustees of Biometrika)

Appendix 3

Present value factors

Years	1%	2%	3%	4%	5%	6%	7%	8%	9%	10%
1	.9901	.9804	.9709	.9615	.9524	.9434	.9346	.9259	.9174	.9091
2	.9803	.9612	.9426	.9246	.9070	.8900	.8734	.8573	.8417	.8264
3	.9706	.9423	.9151	.8890	.8638	.8396	.8163	.7938	.7722	.7513
4	.9610	.9238	.8885	.8548	.8227	.7921	.7629	.7350	.7084	.6830
5	.9515	.9057	.8626	.8219	.7835	.7473	.7130	.6806	.6499	.6209
6	.9420	.8880	.8375	.7903	.7462	.7050	.6663	.6302	.5963	.5645
7	.9327	.8706	.8131	.7599	.7107	.6651	.6227	.5835	.5470	.5132
8	.9235	.8535	.7894	.7307	.6768	.6274	.5820	.5403	.5019	.4665
9	.9143	.8368	.7664	.7026	.6446	.5919	.5439	.5002	.4604	.4241
10	.9053	.8203	.7441	.6756	.6139	.5584	.5083	.4632	.4224	.3855
11	.8963	.8043	.7224	.6496	.5847	.5268	.4751	.4289	.3875	.3805
12	.8874	.7885	.7014	.6246	.5568	.4970	.4440	.3971	.3555	.3186
13	.8787	.7730	.6810	.6006	.5303	.4688	.4150	.3677	.3262	.2897
14	.8700	.7579	.6611	.5775	.5051	.4423	.3878	.3405	.2995	.2633
15	.8613	.7430	.6419	.5553	.4810	.4173	.3624	.3152	.2745	.2394
16	.8528	.7284	.6232	.5339	.4581	.3936	.3387	.2919	.2519	.2176
17	.8444	.7142	.6050	.5134	.4363	.3714	.3166	.2703	.2311	.1978
18	.8360	.7002	.5874	.4936	.4155	.3503	.2959	.2502	.2120	.1799
19	.8277	.6864	.5703	.4746	.3957	.3305	.2765	.2317	.1945	.1635
20	.8195	.6730	.5537	.4564	.3769	.3118	.2584	.2145	.1784	.1486

Years	11%	12%	13%	14%	15%	16%	17%	18%	19%	20%
1	.9009	.8929	.8850	.8772	.8696	.8621	.8547	.8475	.8403	.8333
2	.8116	.7972	.7831	.7695	.7561	.7432	.7305	.7182	.7062	.6944
3	.7312	.7118	.6931	.6750	.6575	.6407	.6244	.6086	.5934	.5787
4	.6587	.6355	.6133	.5921	.5718	.5523	.5337	.5158	.4987	.4823
5	.5935	.5674	.5428	.5194	.4972	.4761	.4561	.4371	.4190	.4019
6	.5346	.5066	.4803	.4556	.4323	.4104	.3898	.3704	.3521	.3349
7	.4817	.4523	.4251	.3996	.3759	.3538	.3332	.3139	.2959	.2791
8	.4339	.4039	.3762	.3506	.3269	.3050	.2848	.2660	.2487	.2326
9	.3909	.3606	.3329	.3075	.2843	.2630	.2434	.2255	.2090	.1938
10	.3522	.3220	.2946	.2679	.2472	.2267	.2080	.1911	.1756	.1615
11	.3173	.2875	.2607	.2366	.2149	.1954	.1778	.1619	.1476	.1346
12	.2855	.2567	.2307	.2076	.1869	.1685	.1520	.1372	.1240	.1122
13	.2575	.2292	.2042	.1821	.1625	.1452	.1299	.1163	.1042	.0935
14	.2320	.2046	.1807	.1597	.1413	.1252	.1110	.0985	.0876	.0779
15	.2090	.1827	.1599	.1401	.1229	.1079	.0949	.0835	.0736	.0649
16	.1883	.1631	.1415	.1229	.1069	.0930	.0811	.0708	.0618	.0541
17	.1696	.1456	.1252	.1078	.0929	.0802	.0693	.0600	.0520	0.451
18	.1528	.1300	.1108	.0946	.0808	.0691	.0592	.0508	.0437	0.376
19	.1377	.1161	.0981	.0829	.0703	.0596	.0506	.0431	.0367	0.313
20	.1240	.1031	.0868	.0728	.0611	.0514	.0433	.0365	.0308	0.261

Appendix 4

Cumulative present value factors

Years	1%	2%	3%	4%	5%	6%	7%	8%	9%	10%
1	.0.990	0.980	0.971	0.962	0.952	0.943	0.935	0.926	0.917	0.909
2	1.970	1.942	1.913	1.886	1.859	1.833	1.808	1.783	1.759	1.736
3	2.941	2.884	2.829	2.775	2.723	2.673	2.624	2.577	2.531	2.487
4	3.902	3.808	3.717	3.630	3.546	3.465	3.387	3.312	3.240	3.170
5	4.853	4.713	4.580	4.452	4.329	4.212	4.100	3.993	3.890	3.791
6	5.795	5.601	5.417	5.242	5.076	4.917	4.767	4.623	4.486	4.355
7	6.728	6.472	6.230	6.002	5.786	5.582	5.389	5.206	5.033	4.868
8	7.652	7.325	7.020	6.733	6.463	6.210	5.971	5.747	5.535	5.335
9	8.566	8.162	7.786	7.435	7.108	6.802	6.515	6.247	5.995	5.759
10	9.471	8.983	8.530	8.111	7.722	7.360	7.024	6.710	6.418	6.145
11	10.368	9.787	9.253	8.760	8.306	7.887	7.499	7.139	6.805	6.495
12	11.255	10.575	9.954	9.385	8.863	8.384	7.943	7.536	7.161	6.814
13	12.134	11.348	10.635	9.986	9.394	8.853	8.358	7.904	7.487	7.103
14	13.004	12.106	11.296	10.563	9.899	9.295	8.745	8.244	7.786	7.367
15	13.865	12.849	11.938	11.118	10.380	9.712	9.108	8.559	8.061	7.606
16	14.718	13.578	12.561	11.652	10.838	10.106	9.447	8.851	8.313	7.824
17	15.562	14.292	13.166	12.166	11.274	10.477	9.763	9.122	8.544	8.022
18	16.398	14.992	13.754	12.659	11.690	10.828	10.059	9.372	8.756	8.201
19	17.226	15.678	14.324	13.134	12.085	11.158	10.336	9.604	8.950	8.365
20	18.046	16.351	14.877	13.500	12.462	11.470	10.594	9.818	9.129	8.514

Year	11%	12%	13%	14%	15%	16%	17%	18%	19%	20%
1	.0.901	0.893	0.885	0.877	0.870	0.862	0.855	0.847	0.840	0.833
2	1.713	1.690	1.668	1.647	1.626	1.605	1.585	1.566	1.547	1.528
3	2.444	2.402	2.361	2.322	2.283	2.246	2.210	2.174	2.140	2.106
4	3.102	3.037	2.974	2.914	2.855	2.798	2.743	2.690	2.639	2.589
5	3.696	3.605	3.517	3.433	3.352	3.274	3.199	3.127	3.058	2.991
6	4.231	4.111	3.998	3.889	3.784	3.685	3.589	3.498	3.410	3.326
7	4.712	4.564	4.423	4.288	4.160	4.039	3.922	3.812	3.706	3.605
8	5.146	4.968	4.799	4.639	4.487	4.344	4.207	4.078	3.954	3.837
9	5.537	5.328	5.132	4.946	4.772	4.607	4.451	4.303	4.103	4.031
10	5.889	5.650	5.426	5.216	5.019	4.833	4.659	4.494	4.339	4.192
11	6.207	5.938	5.687	5.453	5.234	5.029	4.836	4.656	4.486	4.327
12	6.492	6.194	5.918	5.660	5.421	5.197	4.988	4.793	4.611	4.430
13	6.750	6.424	6.122	5.842	5.533	5.342	5.118	4.910	4.715	4.533
14	6.982	6.628	6.302	6.002	5.724	5.468	5.229	5.008	4.802	4.611
15	7.191	6.811	6.462	6.142	5.847	5.575	5.324	5.092	4.876	4.675
16	7.379	6.974	6.604	6.265	5.954	5.668	5.405	5.162	4.938	4.730
17	7.549	7.120	6.729	6.373	6.047	5.749	5.475	5.222	4.990	4.775
18	7.702	7.250	6.840	6.467	6.128	5.818	5.534	5.273	5.033	4.812
19	7.839	7.366	6.938	6.550	6.198	5.877	5.584	5.316	5.070	4.843
20	7.963	7.469	7.025	6.623	6.259	5.929	5.628	5.353	5.101	4.870

Index